THE SCALPEL AND THE BUTTERFLY

THE SCALPEL AND THE BUTTERFLY

THE WAR BETWEEN
ANIMAL RESEARCH AND
ANIMAL PROTECTION

DEBORAH RUDACILLE

Farrar, Straus and Giroux

NEW YORK

FARRAR, STRAUS AND GIROUX
19 Union Square West, New York 10003

Copyright © 2000 by Deborah Rudacille
All rights reserved
Distributed in Canada by Douglas & McIntyre Ltd.
Printed in the United States of America
Designed by Robert C. Olsson
First edition, 2000

Library of Congress Cataloging-in-Publication Data

Rudacille, Deborah.
    The scalpel and the butterfly : the war between animal research and animal
protection / by Deborah Rudacille.—1st ed.
        p.   cm.
    Includes bibliographical references (p.   ).
    ISBN 0-374-25420-6 (alk. paper)
        1. Animal experimentation—Moral and ethical aspects.   2. Animal welfare—
Moral and ethical aspects.   I. Title.

HV4915 .R829 2000
174'.4—dc21   00–028758

For my father

CONTENTS

Power, after investing itself in the body, finds itself exposed to a counter-attack in that same body.

—Michel Foucault, "Body/Power"

THE SCALPEL AND THE BUTTERFLY

INTRODUCTION

Titan! to whose immortal eyes
The sufferings of mortality,
Seen in their sad reality,
Were not as things that gods despise;
What was thy pity's recompense? . . .

Thy Godlike crime was to be kind,
To render with thy precepts less
The sum of human wretchedness
And strengthen Man with his own mind . . .
Thou art a symbol and a sign
To Mortals of their fate and force;
Like Thee, Man is in part divine,
A troubled stream from a pure source . . .

—George Gordon, Lord Byron,
from "Prometheus," 1816

IN 1992, John Orem, professor of physiology at Texas Tech Health Sciences University, was asked by science reporter Ron Kaufman to comment on the Animal Enterprises Protection Act, a law passed by Congress to make vandalism of animal research laboratories a federal crime. Orem, whose laboratory had been trashed by vandals on July 4, 1989, and who was to endure years of harassment by animal rights activists who objected to his sleep research on cats, commented that although FBI involvement in investigating these crimes might aid local police, "the real question to ask in order to stop the ALF [Animal Liberation Front] is what fuels this deep distrust of scientists? What is behind the disaffection that would turn people violently opposed to biomedicine?"

These questions have seldom been asked by those who have sought to defend biomedical research against its critics. But they provide a key to understanding not only the rage of those few individuals who have vandalized laboratories and harassed scientists, but also the uneasiness of many members of the public and the nonscientific establishment who generally support biomedical research, with occasional reservations. To answer these questions we must journey into the past, as far back as 1816, when a young woman with impeccable literary and political bloodlines sat down to write a story on a dare—and provided the world with an enduring image of demonic science.

Geneva, 1816

THE CLOCK strikes two and a woman is writing. Candlelight flickers over the page as she dips her pen into the inkwell. Shadows wrestle on the uneven walls of the villa, and outside a rising wind rustles though the leaves, foreshadowing another torrential rain. A flash of lightning illuminates the room as the woman's sleeping lover stirs in the big curtained bed behind her. The woman writes feverishly, committing her own nightmare to paper. A man, brilliant and obsessed, imagining that he will find a way to overcome death forever. A creature, assembled from disparate parts, a patchwork of corpses, stitched together by the scientist and galvanized into life by an electrical current. The creature's agony and his creator's horror; death not overcome but immeasurably increased.

Mary Godwin Shelley was nineteen years old when she wrote the novel that was to make her name, and that of her creation, immortal. Daughter of two radical eighteenth-century writers, Mary Wollstonecroft and William Godwin, the young woman had eloped with another visionary artist, the poet Percy Shelley, in July 1814. Though still young, Mary had already experienced tragedy, including the death of her mother when she was only a few days old and the loss of her first baby. During the summer of 1816, she and Shelley met and befriended another English poet, George Gordon, Lord Byron, who was renting a neighboring villa in Geneva. Soon after they met, Byron read to Mary, Shelley, and his other guests a newly completed poem about Prometheus, who had stolen fire from the gods and been condemned to eternal torment for his crime. In one version of the Greek myth, Prometheus had offered the sacred fire to human beings, who until then had lived like animals, lacking both heat and light; in another version he had used the divine spark to create man. But in both cases, Prometheus had stolen that which belonged to the gods alone, and for that hubris he was punished.

Many years later, Mary Shelley described the environment in which she had conceived her monstrous progeny. "Many and long were the conversations between Lord Byron and Shelley, to which I was a devout but nearly silent listener," she wrote. "During one of these, various philosophical doctrines were discussed, and among others the nature of the principal of life, and whether there was any probability of its ever being discovered and communicated." Byron's guests, who included his physician, John Polidori, speculated that it might be possible to reanimate a corpse. "Galvanism had given token of such things; perhaps the component parts of a creature might be manufactured, brought together and imbued with vital warmth."

Retiring to bed one night soon after Byron had challenged each of his guests to write a tale of horror, Mary lay sleepless. "I busied myself to think of a story—a story to rival those which had excited us to this task. One which would speak to the mysterious fears of our nature, and awaken thrilling horror." Soon enough, her mind produced a vision that would haunt not only her own generation, but each successive generation—a figure who combined atavastic fears of the corruption of the grave and decay and death with a nascent suspicion of human efforts to overcome these ancient terrors. Over the weeks and months that followed, Mary Shelley poured all of the intellectual intensity of her heritage and social circle into her scientist hero. He was to be the modern Prometheus, and in his name she married brightness with shadow, just as in his character she wove strands of heroism and villainy. She called him Victor, the triumphant one. But the name by which he and his creation would be remembered was his shadowy surname, heavy and Germanic. No doubt she whispered the name to herself as she wrote in the dim room that stormy summer of 1816: Frankenstein.

"I saw—with shut eyes, but acute mental vision—I saw the pale student of unhallowed arts kneeling before the thing he had put together. Frightful must it be; for supremely frightful would be the effect of any human endeavor to mock the stupendous mechanism of the Creator of the world." Mary Shelley wrote those words fifteen years later, in the introduction to the 1831 edition of her novel. Perhaps one of the reasons her creation has flourished and endured, compulsively reworked by other artists, endlessly parodied and imitated and discussed, is that the primal awe and dread expressed so compellingly by Mary Shelley at the dawn of the research era remain inextinguishable components of human psychology today. We retain the primitive sensibilities of our ancestors, one of which is an instinctive shrinking from activities we suspect are taboo. Individuals who violate these taboos, whether by cutting into the dead bodies of humans or the living bodies of animals, have traditionally been viewed with both awe and terror. Science may have rejected the metaphysical tenets of shamanism, but that does not mean that scientists have not assumed the cultural role of the shaman, the hero who ventures into the liminal world between life and death to heal disease and to preserve life and health. Such underworld voyagers have always been simultaneously respected and feared.

"To examine the causes of life, we must first have recourse to death," Shelley's fictional scientist avows. "Darkness had no effect upon my fancy; and a churchyard was to me merely the receptacle of bodies deprived of life, which from being the seat of beauty and strength, had become food for the worm." This bold embrace of materialism, and rejection of any supernatural realms beyond the grave, is coupled in Shelley's novel (as it was in real life) with a new source of terror. "Who shall conceive the horrors of my secret toil, as I dabbled among the unhallowed damps of the grave, or tortured the living animal to animate the lifeless clay," Frankenstein tells Captain Walton, his all but silent listener. "I collected bones from charnel-houses; and disturbed, with profane fingers, the tremendous secrets of the human frame."

In an era when "resurrection men" were removing fresh corpses from graves and selling them to scientists for dissection, and opposition to animal experimentation was related to fear of one's corpse being sold for dissection, Shelley's ghoulish scientist was a new kind of bogey to haunt the dreams of adults and children alike. Over the course of the next two centuries, as biomedical researchers ever more successfully probed "the tremendous secrets of the human frame," scientists' stature grew and they were no longer perceived as villains by most people. Yet somehow, rather than being assuaged by science's greatly expanded understanding of human physiology and increasing power to combat disease and death, certain anxieties have multiplied as the boundaries of the "natural" continually recede and science's powers, the type of powers once termed "occult," increase. The word *Frankenstein* remains a compelling metaphor for many at the turn of the millennium, as we contemplate placing animal organs in human bodies as substitute parts, manipulating the humane genome to treat or prevent disease, and remaking human and animal bodies from the inside out.

"In a culture in which organ transplants, life-extension machinery, microsurgery, and artificial organs have entered everyday medicine, we seem on the verge of practical realization of the seventeenth-century imagination of body as machine," writes postmodern philosopher Susan Bordo. "Western science and technology have now arrived, paradoxically but predictably (for it was an element, though submerged and illicit, in the mechanist conception all along), at a new, postmodern imagination of human freedom from bodily determination. Gradually and surely, a technology that was first aimed at the replacement of malfunctioning parts has generated an industry and an ideology fueled by fantasies of rearranging, transforming, and correcting, an ideology of limitless improvement and change, defying the historicity, the mortality, and indeed the very materiality of the body."

Over one hundred years ago, Charles Darwin showed that humans are a kind of animal. But we are the animal that is not content with the pleasures and limits of animal life. Instead, we push beyond the limits imposed by nature and seek to remake the world, our bodies, and our fates. We seek the powers of gods—and science has given us those powers, producing tremendous benefit and equal anxiety. Scientists, like Prometheus, attempt to acquire knowledge to enhance the lives of human beings. But as the myth of Prometheus and the similar myth of Faust teach us, such knowledge is only acquired at great cost— the cost of one's soul. Whether one believes in the soul as a metaphysical reality or as a shorthand term used to signify an acceptable moral and ethical outlook, for many critics of biomedical research, this loss of "soul" has been clearly evident in the use of animals as the subjects of scientific research and the related objectification and identification of the human person with his or her body.

As the ancient but limited practice of vivisection (experimentation on living animals) was developed into a systematic mode of biological study in the nineteenth century, and knowledge of the body and its workings in health and disease began to grow, critics arose to denounce both the assumptions on which the new methodology was based and the characters of the men wielding the scalpels. Like Victor Frankenstein, these researchers often appeared ghoulish and irreligious to their contemporaries by immersing themselves in matter, oblivious to both the pain they were inflicting on animals and the God-given revulsion their critics thought should prohibit such explorations. Early antivivisectionists contended not only that vivisection was morally wrong but that experiments on animals would inevitably lead to experiments on human beings. Others denied that vivisection served any useful purpose, contending that the practice was simple sadism dressed up in the language of science. The assumption behind all criticism, whether spoken or implicit, was that animal experimentation was "unnatural," that it violated some crucial taboo in a way that even a carnivorous diet did not.

In the nineteenth century, the antivivisection movement in England included some of the most influential and respected members of the British cultural establishment, including the writers Browning, Tennyson, Carlyle, and Ruskin, the great social reformer Lord Shaftesbury, and many other eminent jurists and churchmen. Queen Victoria herself made no secret of her sympathy for the movement. Literary opposition to the practice of animal experimentation even then had a long history; Voltaire and Samuel Johnson and, later, Victor Hugo and George Bernard Shaw were only a few of the writers who spoke out against the practice.

H. G. Wells is an interesting figure to contemplate when considering this topic. Mary and Percy Shelley were prototypical Romantics. But Wells, born over half a century later, was a student of Thomas Huxley and an avid proponent of the benefits of science and technology. His vision of the science-saturated world of the future was often utopian. Yet even this friend of science was compelled to create a scientist hero who is similar in many ways to the complex figure envisioned by Mary Shelley eighty years earlier, and a world in which scientific research unleashed from ethical constraints creates only horror.

In *The Island of Dr. Moreau*, published in 1896, Wells gives full expression to the nebulous anxieties and existential dread of a post-Darwin culture just beginning to grapple with the implications of evolution. The dawning understanding that humans, too, were animals, together with the rapid growth of animal experimentation following the successful experiments of bacteriologists and the development of vaccines for certain infectious diseases shared by humans and animals, created a dilemma. In Moreau's laboratory, animals are painfully vivisected in order to be made nearly human, but as Prenderick, a man cast adrift on the island, soon realizes, the process could be reversed. Humans could be made into animals. Prenderick's horror as he hears the screams of a puma being vivisected echoes the anguish experienced by many people when they are first confronted with the high price of scientific knowledge in animal life and suffering—and the inchoate terror of the human animal that imagines itself an equally helpless experimental subject.

"It was as if all the pain in the world had found a voice," Wells wrote, "yet had I known such pain was in the next room, and had it been dumb, I believe—I have thought it since—I could have stood it well enough. It is when suffering finds a voice and sets our nerves quivering that this pity comes troubling us." Lacking a voice themselves, laboratory animals have been represented by the angry, anguished voices of antivivisectionists and other activists who echo Moreau's experimental subjects in naming the research laboratory a "house of pain." This movement has waxed and waned and waxed again over the past hundred and fifty years, but it has never entirely died out and perhaps never will, because at some level antivivisectionists express a discomfort common to many people when they are confronted with the reality of animal suffering and its roots in experimentation to acquire knowledge of human disease.

Biomedical scientists have traditionally carried out work that many people have found disturbing. Unlike the slaughtering of animals for food, which until two generations ago was part of many people's experience, biological experimentation has always been the province of a self-selected elite. Dissecting dead human beings to understand anatomy and creating disease in animals and cutting them open to discover how the body's organs and systems are affected are perceived as gruesome work by many nonscientists, and no amount of Promethean rhetoric has ever been enough to convince some people that they are justified. Even those willing to grant the necessity of these activities often do not wish to know too much about laboratories or the animals who live and die there. A kind of "don't ask, don't tell" policy has long been in effect, whereby society will permit animal experimentation— and certain types of research on human subjects—as long as it is protected from the details.

Lay criticism and attempts to control research activities are not new phenomena, nor are the determined efforts of scientists to free themselves from public criticism and political control. This conflict has existed from the early years of biomedical research and is not universally based on concerns about animal welfare. What is new is the way in which public and academic suspicion of science has allied itself with the condemnation of the Enlightenment culture that gave birth to both modern science and democracy, colonialism and technology. Christopher Columbus, Isaac Newton, Louis Pasteur, and Thomas Jefferson are now viewed equally, in some quarters, as the bearers of tainted gifts. New worlds were discovered, but at what cost? And wouldn't it be better if human beings had stayed in the old world, where they belonged—the world of natural limits in which new continents were left unspoiled, animals uncaged, indigenous peoples unenslaved, and viruses free to frolic in their hosts, reducing the number of avaricious Westerners infesting the earth?

This is a tongue-in-cheek summary of some of the most extravagant claims of a motley collection of cultural critics, sometimes designated "the academic left" by their adversaries, though not all are academics. The perspectives espoused by various critics of modernity, science, and technology may lend themselves to parody, yet the genuine scholarly contribution to our understanding of both the positive and negative aspects of Enlightenment culture that underly their critique is valid. For too long, only the positive aspects of this legacy were acknowledged, but over the past thirty years, much that was once denied or repressed has come to light and has been hotly debated in the pages of scholarly journals and in the media. Initially, biomedical scientists ignored this type of criticism. Immersed in their own research, many were unaware that the work of various feminist, Afrocentric, or postmodern literary scholars had any relevance at all to their own work, work that they firmly believed was not only useful but also ethically sound.

But as the number of academic papers and books and scholarly journals and articles propounding a perspective that is at best suspicious of science's hegemony grew, a few researchers charged forth from their laboratories to counter the enemy. In *Higher Superstition: The Academic Left and Its Quarrels with Science*, published in 1994, biologist Paul R. Gross and mathematician Norman Levitt state in their preface that "the writing of *Higher Superstition* was undertaken only when it became clear to us, from separate but remarkably similar experiences at our respective universities, that something new and unwelcome had found its way into the academic bloodstream and thence into lecture rooms, journals, books and faculty chit-chat: the systematic disparagement of modern science. A public response was clearly needed."

However valiant their attempts to counter the attacks of disaffected humanists and activists Gross and Levitt failed, like many other scientists before them, to ask and address the simple question expressed so powerfully by John Orem in 1992: "What fuels this deep distrust of scientists?" In the face of the measurable improvements that biomedical research in particular has made, and is making, in improving human life and health—many formerly lethal diseases conquered and life spans increased—"What is behind the disaffection that would turn people violently opposed to biomedicine?"

This book is an attempt to answer that question fairly, with respect to both science and its critics. As a science writer, I am fully aware of the value of biomedical research and the role that animal experimentation has played in enhancing human (and animal) health, and I will attempt to describe a few of those contributions. There can be no doubt that the scientific revolution and its aftermath, the age of scientific medicine in which we live, have vastly improved life for millions of human beings. Any critique of science that does not take into account the astounding advances in human health over the past hundred and fifty years—infectious diseases like smallpox wiped out, once dreaded killers of children like diphtheria and polio conquered, common illnesses easily treated in most cases—must be viewed with suspicion.

However, I also share some of the concerns expressed by those who assert that we have paid, and will continue to pay, a price for this knowledge and that profound ethical and philosophical dilemmas pervade the enterprise. It was not until late in the twentieth century that the rights of either animal or human subjects of research were seriously debated, and even today the nature of those rights in many instances remains a matter of contention. The costs of viewing animals and humans as biological machines and nature as an inexhaustible reservoir of resources for human manipulation and consumption are by now apparent to all but the most opaque observers. Nonetheless, the sweeping condemnation of "Western culture" implicit in revolutionary critiques such as deep ecology, ecofeminism, and animal rights exhibits the same flaws as every other totalizing ethic in human history. Those who embrace a pure vision, whether of a world redeemed by science or one destroyed by it, are blind to ambiguity and unwilling to ask the wrenching questions that accompany each new discovery. At present, particularly in the field of genetic medicine, these questions are especially painful and difficult to answer.

For anyone committed to the democratic tradition in which competing philosophies and practices are permitted to flourish, the debate surrounding the scientific use of animals and its connection to larger issues is bound to be problematic. At times it seems that scientists and those committed to the scientific worldview stand on one side of a great gulf, and many writers, philosophers, and activists on the other. This schism between the humanities and the sciences was identified by C. P. Snow in *The Two Cultures and the Scientific Revolution*, published in 1960. Snow, who was both a scientist and a novelist, noted that scholars in the humanities are, like the public, often ignorant of basic scientific facts, and that those who know little of science are in a very weak position to analyze and critique it. However, it is also true that a great deal of the resentment that one finds directed against science and scientists is linked to the kind of elitist disparagement of lay concerns that permeates science. This tendency to dismiss all criticism as uninformed and unfounded is evident in books like *Higher Superstition*, which attempts to counter challenges posed by fellow academics, and is even more pronounced when the scientific community is confronted by individuals who do not possess academic credentials but nonetheless have legitimate questions about both the means and the ends of research.

At times it appears that it will be impossible to bridge the chasm that separates those who believe that under no circumstances is it permissible for animals to be used in scientific experimentation (or for humans to order the natural world according to their own needs and desires) from those to whom the natural world, including even the human genome, is a plastic and infinitely malleable tool. We may soon have within our grasp the power to remake ourselves, at the most basic level. What might be the outcome of this experiment none can now foresee. But it is certain that history has something to teach us about the dangers of both scientific hubris and public ignorance of science. Somehow these two problems seems a matched set, both a cause of the conflict described in the pages of this book, and its direct result.

A few years after Mary Shelley wrote *Frankenstein, or The Modern Prometheus*, her husband published a work that championed the opposing perspective. In his great poem *Prometheus Unbound*, Percy Bysshe Shelley expressed a hope that he shared with William Godwin: a world set free by the power of science and reason. Mary Shelley was aware of her father's desire for a society freed from ancient superstition and religious orthodoxy, a rational utopia. But in her own masterpiece, she delineated the horrifying outcome when reason is divorced from feeling and science from ethics. She dedicated her book, with its masterful depiction of the pain caused by rationality and intellect unleavened by compassion, to her father. Rather than viewing this family quarrel as a purely personal matter, it may be helpful to see it as one manifestation of an ongoing struggle, one that neither side can ever win but that is nonetheless necessary and appropriate.

"When Shelley pictured science as a modern Prometheus who would wake the world to a wonderful dream of Godwin, he was alas too simple," commented physicist Jacob Bronowski as he contemplated the wreckage of Nagasaki. "But it is as pointless to read what has happened since as a nightmare. Dream or nightmare, we have to live our experience as it is, and we have to live it awake. We live in a world which is penetrated through and through by science, and which is both whole and real. We cannot turn it into a game simply by taking sides."

ONE

VIRUSES, VACCINES, AND VIVISECTION

To learn how men and animals live, we cannot avoid seeing great numbers of them die, because the mechanism of life can be unveiled and proved only by knowledge of the mechanisms of death.

—Claude Bernard,
Introduction to the Study of Experimental Medicine

PARIS, 1876

SLOWLY AND PAINFULLY, professor of physiology Claude Bernard finishes his weekly lecture at the Collège de France. "The experimental medicine which it is my duty to teach you has not yet been born," he concludes, quoting from his much praised textbook, *Introduction to the Study of Experimental Medicine*. "But I promise you that our work will bring forth new hope for the sick and the dying."

The students applaud politely and rise as Paul Bert, Bernard's former pupil, enters the room. Some of them smile and shake their heads. The old man is crazy, of course. How can experiments on dogs and sheep and rabbits help their patients, the children who die by the hundreds when diphtheria strikes, the old people taken by influenza every winter, the unfortunate women who give birth in perfect health but die thrashing in fever a few days later. Sheer lunacy to think that cutting apart animals will help these people. Bernard himself admits that often when he begins an experiment, he has no clear idea of what he is looking for. "The experiment itself supplies me with the key questions," he says.

"Witchcraft," one of the medical students mutters to himself as he hurries off to the hospital where the poor lie crowded together, sometimes two and three to a bed. "What a waste of time when people are sick and dying," he says to a colleague, who nods.

But other students, curious and intense, crowd around Bernard at the lectern and ply him with questions as Paul Bert approaches.

"Professor, pardon, but Pasteur says that you must come to him."

Bernard looks up in surprise.

"It is a very great discovery," Bert says. "Koch has shown that anthrax is caused by a microorganism. He has grown the pathogen in culture and used it to infect an animal. A student has just arrived from Breslau, where Koch demonstrated the procedure to Cohnheim and Weigert."

Without a word, Bernard picks up his satchel and marches ahead of Bert out of the room, while an excited buzz begins to emerge from the crowd of students.

CLAUDE BERNARD was not the first scientist to experiment on an animal, but he was one of the first to provide a cogent account of the scientific rationale for vivisection in his book *Introduction to the Study of Experimental Medicine*, published in France in 1865. Bernard described his work to fellow researchers as well as a general audience, and he contributed a number of articles to the popular press. A lucid and engaging writer, Bernard in his youth had planned to be a playwright until he was dissuaded by a famous French critic who advised him to study medicine instead. His early career change, from dramatist to scientist, presaged the great cultural shift that was to take shape in the latter years of the nineteenth century. Artists, philosophers, and theologians had once stood at the pinnacle of cultural power, forming an elite cadre of "opinion leaders." But by the start of the new century, scientists were to assume this mantle of power. Within the space of a generation, a new priesthood was being founded, the priesthood of science—explicitly articulated as such by the leading proponents of the new gospel and by its critics.

Claude Bernard is significant as well in that his personal life mirrors to a startling degree the vast chasm between those who insisted on the scientist's need and right to experiment on animals and those who objected strongly to the practice of vivisection. Marie-Françoise Bernard abhorred her husband's work and could not understand why he experimented on animals when he could instead have devoted himself to sick people and built a flourishing medical practice. For that reason, Madame Bernard has not been treated kindly by her husband's biographers, who tend to dismiss her as an uneducated woman who made Bernard's home life a hell and deprived him of the company of his daughters. Viewed from a contemporary perspective, Madame Bernard is a rather sad figure whose life reflects the frustrations and limitations endured by women of that era. Deeply sympathetic to animal suffering, Marie-Françoise Bernard not only found herself married to a man who readily engaged in animal experimentation, but also provided unwitting financial support for those activities, via her dowry and comfortable annual income. Most of his biographers agree that Bernard married Marie-Françoise to gain access to her income to support his research activities. The marriage, arranged by a colleague of Bernard's, was ended by formal separation in 1870—a startlingly modern outcome for this nineteenth-century French Catholic couple and a testimony to their mutual unhappiness and fundamental incompatibility.

Marie-Françoise Bernard objected to her husband's work on moral grounds, but he had enemies in the medical profession as well. In the first few decades of the nineteenth century, disease remained a great mystery, believed by many to be caused by "miasmas," or foul odors. The men who locked themselves in basement and attic laboratories to vivisect animals in order to understand the workings of the body in health and sickness were viewed as oddities by the physicians of the time, who considered their work irrelevant at best. Although no one really knew the causes of afflictions such as tuberculosis, diphtheria, typhus, and plague, most of Bernard's medical peers were certain that they were not to be found in the bodies of mutilated animals.

This was true despite the fact that curious human beings had been vivisecting animals since antiquity. Galen of Pergamon, the second-century Roman physician whose work formed the basis of medical practice in the West for more than one thousand years, experimented on pigs, dogs, and other animals. Galen experimented on animals to study anatomy, and this remained the reason for most animal experimentation well into the modern era. For many centuries, the Church forbade the dissection of human bodies, and even after the bishops relented, the public still expressed strong reservations about the practice. Animals provided a handy substitute, and since they were not believed by most people to possess a soul, an experimenter did not endanger his own soul by experimenting on them.

Many early experimenters shared the views expressed by the seventeenth-century French philosopher René Descartes, who believed that animals were a kind of machine. Lacking the ability to reason, expressed in language, and unable to reflect on its own nature and existence, an animal was an automaton, a machine that contains its own principles of motion. The modern word *robot* perhaps best expresses the flavor of the Cartesian view of animals. Unlike Aristotle, who maintained that animals did have a soul, Descartes argued that only man had been granted a soul by the Creator. Lacking consciousness, language, and a soul, animals might react to experimental stimulus, but their reactions were mere reflex and did not necessarily indicate suffering.

"The capacity of animals for sensation, according to Descartes, was strictly corporeal and mechanical, and hence they were unable to feel real pain," National Library of Medicine historian John Parascandola told an audience of researchers and animal protectionists at the first World Congress on Animals and Alternatives in the Life Sciences in 1993. "They just went through the external motions which in man were symptomatic of pain, but did not experience the mental sensation. Some of his followers denied that animals possessed even the inferior kind of feeling that Descartes attributed to beasts, and they interpreted the cries of an animal during vivisection as the mere creaking of the animal 'clockwork.' "

With Bernard's book as founding text, the practice of animal experimentation spread rapidly in the latter decades of the nineteenth century. Young physicians and students from England and the United States visited laboratories in Paris, Breslau, Vienna, Strasbourg, and Berlin to further their medical educations. Most of these students were forever changed by the experience, converted to the new ideal of a scientific medicine. Many years later, William Henry Welch, a pivotal figure in the development of scientific medicine in the United States who had studied in Germany during those years, urged his own students then setting out on their European studies to "get in contact with the great teachers. Then you will have an impression of the men and their work which you will never forget, and every time you read their writings you will remember. Everything will be much more vivid to you."

PARIS, 1879

IN A SMALL ROOM IN PARIS, a man returns from a holiday in the country. He is relaxed and well rested, having left the city in July and returned as the October light cast a soft, nostalgic glow over the city. The man, a chemist named Pasteur who is already famous in France for his work on fermentation, walks over to the shelves pushed up against the wall, shelves that hold three covered flasks filled with a cloudy liquid. Pasteur shakes his head. He had instructed one of his assistants to dispose of the flasks three months ago, before he left for Arbois. The glass vials contain cholera, a killer responsible for the deaths of millions of farm animals. "Idiot," Pasteur grumbles under his breath.

As he busies himself in the laboratory, reading through some notes, he decides to renew the experiment he had been conducting when he left for his holiday in the country. Calling one of his assistants, Pasteur casually lifts a vial of chicken broth culture, which he had infected with the cholera microbe three months before. Poor chickens, he thinks, as he makes his way to the yard behind the building.

A few days later, he sits in the yard, perplexed by the appearance of his hens. They peck and strut in the weak October sun happily, as if he had not just a few days earlier injected them with the cholera-laden broth. Here was a puzzle. What was it about these particular chickens that rendered them immune to cholera? While he ponders this conundrum, Pasteur instructs one of his assistants to prepare a fresh culture of chicken broth teeming with cholera microbes. Later that afternoon they inject the miracle chickens once again. Over the next few days they carefully monitor the birds for signs of sickness. Although the chickens seem a bit less energetic than usual, they show no signs of cholera. Staring at the healthy chickens, Pasteur is baffled. Then an idea dawns, a crazy, wonderful idea. Pasteur rushes into the laboratory, calling all of his students and assistants together.

"We need more chickens," he says.

BY THE TIME he discovered that an injection of weakened cholera microbes will protect animals from developing the full-blown disease, Louis Pasteur was already one of the most famous scientists of the age. During the 1860s, he had performed an invaluable service for the French government by discovering that heat kills the microbes that cause fermenting wine to turn bitter and undrinkable. A few years later, he found the microbe that attacks silkworm eggs, causing a disease that kills large numbers of the animals. These discoveries naturally saved the government, and French silk farmers and winegrowers, hundreds of thousands of francs and made Pasteur a national hero. Unlike his recently deceased rival Claude Bernard, Pasteur was also happily married and enjoyed the unconditional support of his wife. Though Bernard had been recognized as a great man by his fellow scientists and had received many honors and awards before his death in 1878, he had never benefited from the public acclaim that was showered on Pasteur. Many speculated that Bernard's death, putatively caused by a mysterious abdominal affliction that neither he nor his fellow physicians could diagnose, was the result of years spent laboring in damp and poorly ventilated basement laboratories. Pasteur, it was rumored, was soon to be granted an entire research institute devoted to microbiological studies, a new building built by the government and equipped with the most modern and efficient laboratory facilities. Despite all this success, Pasteur was as maligned by antivivisectionists as Bernard had been, although it was said that he was far more sensitive to animal suffering and could not bear to be present during certain procedures. His next round of experiments were to make him even more unpopular with those who objected to experiments on animals.

In May 1881, Pasteur vaccinated twenty-five sheep, six cows, and one goat with a series of inoculations. The first vaccine, administered on May 5, contained a very weak anthrax culture. On May 17, Pasteur inoculated the animals with a more potent, although still attenuated (weak), virus. Another twenty-five sheep, four cows, and one goat were used as controls; they received no vaccinations. On May 31, Pasteur injected all sixty-two animals with an extremely virulent culture of anthrax. Two days later, all of the vaccinated animals but one sheep were well, and all of the control animals dead or dying.

Four years later, Pasteur and his colleagues were working on rabies. They couldn't grow the virus in culture, so they injected rabies-infected cow brain into the heads of rabbits. They grew the rabies in one rabbit and then killed the animal, injecting tissue from its rabies-infected brain into another rabbit's brain. After fifty passages (rabbit-to-rabbit injections) to fix the incubation period of the virus at seven days, the investigators killed the animals and removed their spinal cords, hanging the cords out to dry in the laboratory. Then they injected dogs with the weakened virus of the spinal cord, waited a few days, and then challenged the dogs with a "hot" rabies virus. The dogs resisted the virus and remained well.

Suddenly, Pasteur was presented with an electrifying opportunity. On July 6, a nine-year-old boy named Joseph Meister who had been bitten by a rabid dog arrived in Paris. The boy's parents, knowing that his death was inevitable, agreed to a series of thirteen injections of Pasteur's vaccine over a ten-day period. This was a terrifying gamble. Pasteur had never inoculated an animal against rabies after it had already been exposed to the virus, and he had never inoculated a human being. But the gamble paid off and the child lived. Laboratories began to buzz with the news of Pasteur's discoveries. Some physicians and scientists attacked both the man and his ideas. But many others followed his lead. A new idea was beginning to take hold, fed by Pasteur's work in Paris and that of his rival, Koch, in Germany—the conviction that disease was caused by invisible pathogens called germs. The experimentalists worked feverishly, searching for the invaders and for vaccines to counter their effects.

In England, a different sort of fever had taken hold. As news of the research spread and the methods used to obtain vaccines became more widely known, angry voices were heard denouncing the unnatural experiments with animals and demanding that they cease. The anger built on old fears and grievances. Nearly one hundred years before, in 1796, an English country doctor, Edward Jenner, had confirmed that cowpox, a mild infection transmitted from cows to diary workers, conferred protection from smallpox, the ghastly disease that had killed and disfigured millions of human beings in regular epidemics. Smallpox was known and feared in the ancient world and Europe long before it traveled to the Americas and nearly wiped out the indigenous peoples of the New World.

Jenner's vaccine of cowpox, or vaccinia virus, appeared to provide protection against smallpox. But editorials denouncing the practice appeared in British newspapers, and cartoons depicting men with cow's heads mocked Jenner and his discovery. A few decades later, near riots had broken out in London in 1824 when Claude Bernard's teacher, François Magendie, had held public demonstrations of animal experimentation in the days before ether was discovered to anesthetize man and beast. The man tortured animals in public, slicing into living flesh as if it were a piece of mutton, as the bound beasts screamed in agony. Men became sick to their stomachs and women fainted. British citizens were revolted by Magendie and his experiments. The French professor and his filthy "investigations" were driven back across the Channel, where less than fifty years before, French citizens had watched their friends and neighbors beheaded, lunching *en famille* on Sunday afternoons as though at a picnic. A barbarous race, the British agreed.

Yet the continental disease of vivisection had infected British laboratories as well. The work had always been carried out, albeit on a small scale and quietly. It was well known in England that William Harvey would not have made his famous discovery of blood circulation in 1628 had it not been for his experiments on live deer. As early as 1831, the British physiologist Marshall Hall proposed five principles that he thought ought to guide research on animals. These principles, formulated by a scientist who had publicly admitted that "every experiment, every new or unusual situation of such being, is necessarily attended by pain or suffering of a bodily or mental kind," did not endear him to his fellow physiologists, who ignored Hall's proposal.

One thing was certain, the British public agreed in 1874. Scientists on their side of the Channel may have conducted animal experiments when necessity drove them to it, but they did not glory in vivisection like the French and the Germans. There was no wholesale torture and slaughter of animals in British laboratories—at least not until the French physiologist Claude Bernard published his infernal text and British scientists began visiting the Frenchman's laboratory and aping his procedures and those of his mentor Magendie.

"Revolting." That's how an eyewitness, a retired officer in the British Navy, described Bernard's experiments in a letter published in 1875 to the editor of a London newspaper. "In that laboratory we sacrificed daily from one to three dogs, besides rabbits and other animals, and after four months experience I am of the opinion that not one of those experiments on animals was justified or necessary," Dr. George Hoggan wrote to the *Morning Post.* "During three campaigns I have witnessed many harsh sights, but I think the saddest sight I ever witnessed was when the dogs were brought up from the cellar to the laboratory for sacrifice. Instead of appearing pleased with the change from darkness to light, they seemed seized with horror as soon as they smelt the air of the place."

Despite such protests, English scientists increasingly practiced vivisection and taught it in the universities. When John Burdon Sanderson, a professor of physiology at London University and Oxford, and a proponent of continental-style vivisection, published an English-language text on experimental physiology in 1873, the steaming kettle of public outrage in Britain began to boil. Burdon Sanderson's *Handbook for the Physiological Laboratory* was a beginner's guide to torture, the antivivisectionists said. The book made no mention of anesthesia, despite the fact that Burdon Sanderson himself had helped develop the official policy of the British Association for the Advancement of Science, which required the use of anesthetics in animal experiments. If the authors of the recommendation ignored it, what must other experimenters be doing? Something must be done, said the antivivisectionist leaders. The people agreed.

ENGLAND PROVED to be a particularly fruitful site for the growth of a powerful, organized antivivisection movement. Although antivivisection sentiment existed on both sides of the Channel and antivivisection societies were formed throughout Europe in the final decades of the nineteenth century, only in England did the movement reach a position of relative power and influence. Why was this so, when conditions in other nations seemed equally ripe for such a development?

For example, Germany, like England, was home to a Romantic tradition that idealized nature and the place of animals in nature. Like Queen Victoria, Bismarck was known to be sympathetic to the antivivisection cause. If England had great artists who propounded a doctrine of purification through vegetarianism and animal protection, so too did Germany—the composer Richard Wagner, who died in 1883, was a fervent and eloquent antivivisectionist. Even in France, home of Descartes and his doctrine of animal machines, Voltaire in the eighteenth century and Victor Hugo in the nineteenth argued forcefully that animals ought not to be used as scientific subjects. And at least one great French researcher, Pasteur, was known by his colleagues to be somewhat squeamish about vivisection, as were such British scientists as Darwin and Huxley, who defended the need for animal experimentation at the same time that they appeared uninterested in conducting such experiments themselves. So why did antivivisectionists succeed in passing a law regulating the practice only in England, while in France, Germany, and the United States scientists either did not encounter such strong organized resistance or were able to successfully overcome it?

The answer to this question is complex, but at least part of the answer must lie in a preexisting philosophical tradition in England that seriously addressed the notion of animal rights, and a legislative framework that by 1821 had already admitted the need for some form of animal protection. In 1780, the British utilitarian philosopher Jeremy Bentham published *An Introduction to the Principles of Morals and Legislation*, a book that neatly answered Descartes's dismissal of animals as mere automatons. Unlike the French philosopher, who found rationality expressed in language the hallmark of a morally significant subject, Bentham said, in a phrase that was to serve as a rallying cry for the modern animal protection movement, "The question is not, Can they reason? nor Can they talk? but, Can they *suffer*?"

This crucial distinction, which shifts the focus of moral significance from the ability to reason to the ability to feel, may at least partially explain the gulf that has often existed between scientists and artists on this matter, and a similar statistical discrepancy between men and women that has persisted into the modern era. Any group that tends to highly esteem "feeling" as a guide to morals and behavior will no doubt follow Bentham in granting animals a higher value than those who find "reason" to be the final arbiter of conscience. However, in nineteenth-century Britain, both feeling and reason argued convincingly for the need for some form of protection for animals.

The first attempts failed. In 1800, Sir William Pulteney proposed a bill outlawing the sport of bullbaiting, a popular working-class amusement. Parliament voted down the bill, with those opposed arguing that "the abolition of bull-baiting interfered with the amusement of the people." In 1809 and again in 1811, the Lord High Chancellor of England attempted to pass bills preventing "wanton and malicious cruelty to animals" and was again successfully opposed by those who argued that "no reason can be assigned for the interference of legislation in the protection of animals unless their protection be connected either directly or remotely with some advantage to man."

However, in 1821, an Irish deputy introduced in the House of Commons a bill for the protection of horses and cattle. Known as Martin's Act, the bill finally passed in 1822 and is generally agreed to be the first modern law protecting animals. This legislation was followed by the founding in London in 1824 of a Society for the Prevention of Cruelty to Animals. As early as 1827, the SPCA issued a call for the regulation of animal experimentation, although at that time the number of experimenters in England was few. This proposal was not implemented.

By the middle of the nineteenth century, the society had a new benefactor, Queen Victoria, and was called the Royal Society for the Prevention of Cruelty to Animals. At the same time, the organization began to take a more active role in condemning animal experimentation, creating a delegation to meet with Emperor Napoleon III in 1861 to protest the vivisection of horses at French veterinary schools. Thirteen years later, in 1874, the RSPCA sued four physicians in Norwich, England, for animal cruelty following an experimental demonstration of the effects of alcohol. The physicians had induced an epileptic fit in a dog by injecting it with absinthe.

This incident, together with the publication of Burdon Sanderson's *Handbook for the Physiological Laboratory* in 1873, led to a great deal of discussion in both scientific journals and the popular press about the utility of animal experimentation and the circumstances under which experiments should be carried out. Burdon Sanderson, widely considered to be the best physiologist in England at the time, had helped draft guidelines on animal experimentation for the British Association for the Advancement of Science in 1871. These guidelines explicitly stated that whenever possible, experiments should be conducted under anesthesia and that all teaching demonstrations on living animals ought to be painless or utilize anesthesia. However, many of the experiments described in Burdon Sanderson's text did not include information on anesthesia, leading readers to conclude that scientists were not following the guidelines.

The stage was set for a confrontation.

TWO

THE KINGDOM OF
THE SPIRIT

I went in my sleep last night from one torture-chamber to another in the underground vaults of a vivisector's laboratory, and in all were men at work lacerating, dissecting and burning the living flesh of their victims. But these were no longer mere horses or dogs or rabbits, for in each I saw a human shape, the shape of a man, with limbs and lineaments resembling those of their tormentors, hidden within the outward form. And I cried aloud, "Wretches! You are torturing an unborn man!" But they only mocked me, for with their eyes they could not see that which I saw.

—Anna Kingsford,
diary entry, February 2, 1880

PARIS, 1875

SOMETIMES IT SEEMED to Anna Kingsford that the city of Paris was an enormous abattoir. In every street, in every home and shop and laboratory, the dreadful slaughter was carried out, thousands of animals killed every day for food, for fashion, for science. If she weren't in Paris on a mission to save the suffering bodies of animals and the souls of human beings corrupted by the twin evils of meat eating and vivisection, she would flee this bloody city and return to her country home across the Channel, in the green heart of England. But Anna knew that she had important work to do here, gaining the medical degree that would qualify her to speak with authority on the four topics she had been been chosen by the Spirit to address: compassion for animals, a vegetarian diet, the empowerment of women, and moral development through purification of the body.

Today, she had a special task to undertake. She hurried forward, her student satchel slung over her arm. Accidentally, she brushed against a man walking in the opposite direction and he began to abuse her until he noticed her pretty face and slender form. Then he murmured a suggestion, which she ignored. These verbal assaults occurred all too often when she attempted to traverse the city alone. But her companion and friend Edward Maitland was not always able to accompany her, and her husband, Algernon Kingsford, was consumed by his duties as a pastor and parent. A clergyman in the Anglican Church, he was caring for Eadith, their daughter, alone in England while Anna pursued her studies abroad. It was an unorthodox arrangement that shocked the conventional, but it suited Anna, Maitland, and Algernon.

Maitland had encouraged her to carry out today's assignment—to publicly question her *chef*, the chief surgeon of the hospital to which she was attached as a student, about the scientific inefficacy of vivisection. Kingsford shuddered at her audacity in even contemplating addressing this question to the *chef* in a public setting, on the wards as she and her fellow students followed him from bed to bed.

"I should feel like a common sailor interrogating the admiral of the fleet on the quarterdeck," she had protested.

But Edward, her dear friend and colleague, had offered gentle encouragement. He reminded her of the great labor they had been assigned, the work that had brought them in a chaste and Spirit-blessed partnership to Paris, the home of the gospel of Materialism, which worshiped the body and raised its needs and its lusts above the thirst for the Spirit. Anna nodded as he spoke, she who had never once in all her years of study experimented on an animal, moving from tutor to tutor before finding one who was willing to honor her delicacy in this area. She passed all of her courses nonetheless and could not understand her fellow students who often converted their attic and basement lodgings into Sunday laboratories, carrying out experiments they said helped them understand the lessons of the week. The thought sickened Anna Kingsford. Too often, while studying in the library, she had heard the dogs nearby howling and whimpering under the hands of an experimenter, or while passing a basement laboratory on the street as the door opened like the gates of hell itself, beheld the awful sights, sounds, and smells of vivisection.

Therefore she steeled herself for the task at hand, encouraged by Edward. "Your position is not that of the other students," he had reminded her. "You are English, eccentric by right." Here he smiled. "Moreover, you are a woman, and then again a very beautiful woman, and finally you have admitted that he likes you very much and respects your abilities. Will he not admire you even more when he sees your courage in addressing him on this subject?"

Swayed by Maitland's arguments, Anna agreed, and together they formulated a question that put the matter quite succinctly: "Why is vivisection insisted upon when, as a method, it is considered unscientific and the conclusions to which it points are rejected as unsound?"

Anna had this latter on good authority. A few days earlier a fellow student had been reprimanded during an oral examination for answering "animal experiment" when the *professeur* asked the best method to test poisons and other drugs. "Then you would employ a method fit for only idle and inaccurate men," the *professeur* had replied sharply, echoing the opinion of many of the older faculty. Yet the new science of physiology was gaining many adherents, even in England, Anna had heard. "They will torture animals, but they will not admit a woman to their schools of medicine," she thought wryly.

Later that afternoon, as her class stood by the bedside of a ten-year-old boy who was deaf, mute, and suffering from an abscess in his shoulder, she took a deep breath and addressed her question to the *chef*. He looked up bemusedly while probing the child's smelly wound and graciously replied, "*Oui, madame. Je voudrais répondre à votre question ce soir, quand le grand assem-*

blage des étudiants convergera." He would address the question that evening before the full assembly of students. Anna was ecstatic. She prayed throughout the afternoon.

When the surgeon ascended the podium, noting that one of their number, a student like themselves (here he turned and bowed gallantly to Anna), had posed an important question regarding the efficacy of vivisection, the room grew hushed. Anna's heart beat hard within her chest and she took deep breaths to calm herself, wary of the exaltation that sometimes crept upon her and caused her to enter the shadowy realm where visions replaced her everyday awareness. To steady herself, she took notes, which she read to Maitland later that night.

"Speaking for myself and my brethren of the Faculté, I do not mean to say that we claim for that method of investigation that it has been of any practical utility to medical science or that we expect it to do so," the surgeon began. "But it is necessary as a protest on behalf of the independence of science as against interference by clerics and moralists. When all the world has reached the high intellectual level of France, and no longer believes in God, the soul, moral responsibility, or any nonsense of that kind, but makes practical utility the only rule of conduct, then, and not until then, can science afford to dispense with vivisection."

Anna was not shocked by the *chef*'s statement, for she had heard this and worse from other professors and students. She was surprised and delighted (as was Maitland) by the surgeon's frankness in expressing the new scientific gospel of Materialism. It would be a potent weapon in their hands when they returned to England and began to lecture and write on the horror that awaited the world if Science were to triumph over Spirit.

ANNA BONUS was born in Stratford, England, in 1846. A sensitive, imaginative child, she published her first poem at the age of nine and continued to write and publish throughout her adolescence and adulthood. As a young woman, she greatly enjoyed riding and the excitement of fox hunting until one afternoon when she had a vision of herself in the place of the fox. She married her cousin Algernon Kingsford in 1867, when she was twenty-one, and gave birth to their daughter the next year. Although her husband was ordained an Anglican minister soon after their marriage, Anna converted to Roman Catholicism in 1872. Soon after her conversion the prosperous Kingsford, who had been left a personal income of seven hundred pounds a year by her father, bought *The Lady's Own Paper, A Journal of Progress, Taste and Art.* Her editorial responsibilities brought her into contact with the journalist, feminist, and social crusader Frances Powers Cobbe.

Cobbe, an Irishwoman twenty-four years older than her editor and far more savvy in the ways of the world, contributed an article to *The Lady's Own Paper* on vivisection, awakening Kingsford's own interest in the subject. The same year that Cobbe's article appeared, Kingsford became a vegetarian. Her vegetarianism was a direct outgrowth of her spiritual beliefs. Unlike many other antivivisectionists and animal welfare advocates of the time, Kingsford believed that consuming animal flesh was morally wrong and that vegetarianism was "the only effectual means to the world's redemption." Cobbe, although an ardent antivivisectionist, never shared Kingsford's interest in vegetarianism. Indeed, according to historian Richard French, author of *Antivivisection and Medical Science in Victorian Society*, Cobbe and many other antivivisectionists were "outrightly hostile to vegetarianism." French explains the discrepancy by noting that "vegetarianism was a movement explicitly dedicated to a reconstitution of society and lifestyles, and antivivisection was first and foremost an agitation with a limited practical objective."

In 1873, Kingsford met the man who was to become her lifelong friend and collaborator, Edward Maitland, a widower, son of a minister, and fellow writer twenty-two years her senior. The partnership of Maitland and Kingsford was based on a shared spiritual doctrine, one that rejected materialism—which they conceived as the worship of the body and the material reality of life on earth—and embraced the life of the spirit. Their friendship was approved by Kingsford's husband, who urged Maitland to accompany Kingsford when she decided to abandon journalism and travel to Paris to study medicine. "At present I am studying medicine with the view of ultimately entering the profession—not for the sake of practice, but for scientific purposes," she wrote to Maitland in August 1873, soon after they met. From the start, Kingsford's interest in medicine was explicitly linked to her desire to pursue "the work of advocacy and redemption" that she felt divinely ordained to carry out.

In Paris, women were permitted to study medicine, although they were not truly welcomed. In *The Life and Times of Anna Kingsford: Her Diaries, Letters, and Work*, a two-volume biography published by Maitland after her death, the resistance of the medical faculty and her fellow students (both male and female) to Anna's enrollment in the spring of 1874 is vividly described.

"I wish I could write in a happier strain," Kingsford wrote to her husband in 1874. "Things are not going well with me. My *chef* at the *Charité* strongly disapproves of women students and took this means of showing it. About a hundred men (no women except myself) went round the wards today, and when we were all assembled before him to have our names written down, he called and named all the students except me, and then closed the book. I stood forward upon this, and said quietly, '*Et moi aussi, monsieur.*' He turned on me sharply, and cried, '*Vous, vous n'êtes ni homme ni femme; je ne veux pas inscrire vôtre nom.*' I stood silent in the midst of a dead silence."

Still, by virtue of her wit, persistence, and intelligence, Kingsford managed to win over her critics and to complete her medical studies. Although she was able to surmount the challenges she faced as a medical student in Paris, she remained in agony over her powerlessness to relieve the suffering of the animals used in teaching and research throughout the city. "I have found my Hell here in the *Faculté de Médecine* of Paris, a Hell more real and awful than any I have yet met with elsewhere, and one that fulfills all the dreams of the mediaeval monks," she wrote on August 20, 1879. "The idea that it was so came strongly upon me one day when I was sitting in the Musée of the school, with my head in my hands, trying vainly to shut out of my ears the piteous shrieks and cries which floated incessantly towards me up the private staircase where Beclard, Vulpian and other devils were torment-ing their innocent victims. Every now and then, as a scream more heart-rending than the rest reached me, the moisture burst out on my forehead and on the palms of my hands, and I prayed, 'Oh God, take me out of this Hell; do not suffer me to remain in this awful place.' "

Kingsford was unable to comprehend the experimentalists' indifference to animal suffering. Claude Bernard's assertion in his famous text that "the physiologist is not an ordinary man: he is a scientist, possessed and absorbed by the scientific idea he pur-sues. He does not hear the cries of animals, he does not see their flowing blood, he sees nothing but his idea, and is aware of noth-ing but an organism that conceals from him the problem he is seeking to resolve" seemed to Kingsford a monstrous heresy.

Her views were shared by a growing number of her country-men, many of them influenced by Darwin's ideas, which appeared to provide a scientific rationale for animal protection by demon-strating the kinship between man and beast. Like Claude Bernard, Charles Darwin had not set out to be a scientist at all. He had planned to become a clergyman, a fact that has often been used to explain his long delay in publishing the findings that provided such a radical challenge to the theology of the day. In his revolu-tionary works, *On the Origin of Species by Means of Natural Selec-tion, or the Preservation of Man and Favored Races in the Struggle for Life* (1859), *The Descent of Man and Selection in Relation to Sex* (1871), and *The Expression of the Emotions in Man and Ani-mals* (1872), Darwin set down an account of the human relation-ship to nature and other animals that shattered popular beliefs about the unique biological and spiritual status of human beings. It is impossible to overstate the scientific and cultural significance of Darwin's work, particularly its open acceptance of biological materialism—the belief that matter is the stuff of life and all men-tal and spiritual activities its effects, and the corollary belief that natural selection is a random process, aiming toward no end and directed by no Mover.

Although the essentially atheistic and materialistic aspects of Darwin's theory were anathema to Anna Kingsford and to many other eminent Victorians, the same people had little difficulty accepting the idea that humans were kin to other animals. Activists working to improve the human treatment of animals had been attempting to make the same point for nearly one hundred years. "If the walls came tumbling down at the sound of Darwin's trumpet, it was because the humble termites had nibbled for years," according to James Turner, author of *Reckoning with the Beast: Animals, Pain and Humanity in the Victorian Mind,* a study of the roots of modern animal protection. If animals were our relatives, however distant, the problem of pain identified by Jeremy Bentham in 1780 assumed an even greater urgency—and nowhere was the problem of animal pain more graphically represented than in the practice of vivisection.

As noted by Turner, the Victorian era was distinguished by an increasing revulsion toward physical pain in both humans and animals. People had once accepted physical pain as a given. From toothaches to tumors, public executions to private beatings of servants and family members—pain was everywhere, and therefore nowhere in particular. "Suffering may be everywhere, but if it is not brought to cultural consciousness, in effect it has no existence—that is, there is no knowledge of it," commented Judith Perkins in a study of the iconography of pain. But by the late nineteenth century, suffering was becoming visible: the suffering of slaves, the suffering of the poor, the suffering of animals, and the suffering of the sick and wounded. "The dread of pain forms one of the most basic bonds uniting us with our ancestors a century ago, a bond overlooked only because it is so obvious," says Turner. "Among the devils haunting the Victorians, as in our own hells, pain was an archdemon."

Anna Kingsford journeyed to Paris to confront and vanquish this demon. "Oh God, take me out of this Hell; do not suffer me to remain in the awful place," she wrote in her diary. "And immediately there came to me, like an answer, these words—'He descended into Hell.' And I felt sure that this is my Hell, and that when I have passed its hateful doors I shall be able to give myself freely and effectively to the work of advocacy and redemption which I so ardently long to begin."

LONDON, 1876

FRANCES POWERS COBBE paced in her study and muttered angrily to herself. A big woman, her tread shook the room as she stomped from one end of the Persian carpet to the other. Her friend and companion, the sculptor Mary Charlotte Lloyd, poked her head in, but Cobbe waved her away. Mary knew better than to disturb her at a time like this, when all she had worked for was threatened.

Cobbe sat down heavily on a velvet armchair and groaned. Over the past few years she had built a firm coalition of allies in the highest levels of government, the Church, the literary world, and medicine—influential people convinced that vivisection was an unnecessary evil, a blot on the moral order. By God, she once believed she had Darwin enlisted in the Cause. She had a letter in which the great man stated that he himself could not vivisect and that he deplored animal experimentation carried out "for mere damnable detestable curiosity." But the vivisectors, led by Huxley and Burdon Sanderson, had carried the day, insisting that scientific progress depended on scientific freedom, and Darwin had cast his lot with his fellow scientists, circulating a petition to protect the rights of experimenters as Parliament debated the suppression of the practice.

Still, even without Darwin and with the weight of the entire British scientific establishment against them, the antivivisectionists had succeeded in persuading the Queen to appoint a Royal Commission to study the question. Their legislative ally, Lord Carnarvon, had then presented a bill in the House of Lords banning experiments on dogs and cats and strictly regulating all other experiments. The bill had come within a hair of passing. Only the death of Carnarvon's mother and his absence from London had given the vivisectors time to regroup and counter Carnarvon's bill with their own weakened version, which set up a system of licensing and regulation but which failed to condemn or even to restrict the practice.

Despite the failure of Carnarvon's bill, such testimony before the Royal Commission as that of the Prussian scientist Emanuel Klein, one of the authors of Burdon Sanderson's *Handbook for the Physiological Laboratory*, actually helped the antivivisectionist cause. Although the guidelines drawn up a few years before by the British Association for the Advancement of Science expressly mandated the use of anesthetics in painful experiments, Klein blithely admitted that he seldom took the trouble to anesthetize his experimental animals.

"Except for teaching purposes, for demonstration, I never use anesthetics where it is not necessary for convenience—as for example with dogs and cats," Klein stated. "On frogs and the lower animals I never use them at all."

The chairman of the commission was incredulous.

"When you say that you only use them for convenience' sake, do you mean that you have no regard at all to the sufferings of the animals?"

"No regard at all," Klein responded imperturbably.

"You are prepared to establish that as a principle which you approve?"

"An experimenter has no time, so to speak, for thinking about what the animal will feel or suffer," Klein answered honestly if tactlessly. "His only purpose is to perform the experiment, to learn as much from it as possible, and to do it as quickly as possible."

Upon further questioning, Klein testified that he believed this perspective was shared by the bulk of English physiologists, if not by the common people. He created an uproar within the hearing room and, within hours, among the press and people with his final assertion, comparing the moral indifference of the experimenter to that of the hunter and the chef.

"Just as little as a sportsman or a cook goes enquiring into the detail of the whole business, while the sportsman is hunting or the cook putting a lobster into a pot of boiling water, just as little can the physiologist be expected to devote time and thought to enquiring what this animal may feel while he is doing the experiment."

Cobbe smiled with satisfaction, remembering Klein's testimony. The Prussian's indifference to animal suffering had disgusted even his fellow scientists and virtually ensured passage of the Cruelty to Animals Act of 1876, the world's first law regulating animal experimentation. Although she was disgusted by the leniency of the Act, Cobbe felt certain that the appointment of a second Royal Commission could be achieved in due time if pressure were brought to bear by her coalition. Mentally, she listed the stellar names of her allies and supporters, luminaries such as the Roman Catholic cardinal Manning, Lord Shaftesbury, Lord Chief Justice Coleridge, Dr. Ellicott, and the Bishop of Gloucester. That March they had founded the world's first society exclusively devoted to antivivisection, the Victoria Street Society for the Protection of Animals Liable to Vivisection. No longer would they attempt to reason with the scientists. Reason had failed. From now on, Cobbe and the members of her coalition would fight for outright abolition.

FRANCES POWERS COBBE was born in 1822 near Dublin, Ireland. In many ways, she and Anna Kingsford might have been natural allies. She too had undergone an extensive period of spiritual searching, rejecting the comfortable Protestant doctrines of her youth, and claimed to be led by a phrase carved on a chair in her grandfather's study—"Deliver him that is oppressed from the hand of the adversary." However, while Kingsford rejected the material world and yearned for a wholly spiritual communion, Cobbe was a pragmatist, committed to political action. Still, like Kingsford, she believed in a God of absolute goodness, holiness, and love, one who would not countenance cruelty.

After her early spiritual crisis and a later illness, during which she felt much abused by the medical profession, Cobbe's views on the nature of faith and science, and the proper relations between the two, were set, as historian James Turner commented in his study of the roots of Victorian animal protection, *Reckoning with the Beast*. "Miss Cobbe had painfully reconstructed a faith to give meaning to her life, but her crisis left two ineradicable marks on her worldview. She now firmly subordinated science to faith as a source of final knowledge and as a guide to conduct; for the rest of her life she remained suspicious of the claims of science and scientists. And she refused to contemplate the possibility of any blemish on the perfect goodness of God and his world; any intimation that God might prove good to some of his creatures and cruel to others put too serious a strain on her hard-won faith. She had surmounted her crisis by grinding down science and its supposedly cruel, godless version of nature. In this she resembled many of her contemporaries."

Cobbe traveled extensively after the death of her parents, and on her return to England, she moved to Bristol, where she worked with the social reformer Mary Carpenter. Laboring in the slums of Bristol in Carpenter's reformatory and "ragged schools" for indigent young people, Cobbe planned to spend her life among the poor. But a leg injury and extended period of convalescence changed all that and led her to journalism. Cobbe wrote on many topics, including the emancipation of women, but the issue that was finally to seize her and to which she would devote her considerable drive and ambition was antivivisection. "Round and fat as a Turkish sultana, with yellow hair, and face mature and pulpy, she could have passed more easily as a jovial grandmother than as a saboteur of science," Turner wrote. "But her hearty plumpness belied a keen mind and one of the deftest polemical pens in Victorian England."

In 1863, articles began to appear in English newspapers describing the vivisection of horses in France, particularly at the veterinary school at Alfort. French veterinary students were using live unanesthetized animals to practice surgery. British animal lovers, among them Miss Cobbe, were horrified. Five hundred veterinarians in England signed a petition requesting that the French use only dead horses for student projects. Miss Cobbe herself wrote an essay titled "The Rights of Man and the Claims of Brutes," for *Fraser's Magazine*, a popular periodical. The article was later reprinted in a scholarly journal, *Studies Ethical and Social.*

Later that year, on a trip to Florence, Cobbe and a group of other English expatriates launched a crusade against Professor Moritz Schiff, who was conducting experiments there. Schiff ignored the activists, despite the fact that many of the 783 signatories to a November 1863 petition circulated by Cobbe were high-ranking Florentine aristocrats. After returning to England, Cobbe and her friend Mary Charlotte Lloyd took a house in Hereford Square and began a comfortable existence, with Cobbe writing for a London paper and the two entertaining a series of illustrious guests, including the scientists Sir Charles Lyell and Darwin.

In her autobiography, *Life of Frances Powers Cobbe as Told by Herself*, Cobbe describes Lyell as "the Man of Science as he was of old; devout, and yet entirely free-thinking in the true sense; filled with an admiring, almost adoring love for nature, and also (all the more for that enthusiasm), simple and fresh-hearted as a child." Cobbe's admiring description of this true scientist of the old school, whose theories were critical in helping Darwin formulate the theory of evolution, is complemented by an anecdote in which Lyell explains Darwin's ideas to Miss Cobbe's guests in a manner both amusing and effective.

"We had been discussing Evolution," Miss Cobbe wrote, "and some of us had betrayed the impression that the doctrine (which he had then recently adopted) involved always the survival of the *best*, as well as the *fittest*. Sir Charles left the room and went downstairs, but suddenly rushed back into the drawing room, and said to me all in a breath, standing on the rug: "I'll explain it to you in one minute! Suppose *you* had been living in Spain three hundred years ago, and had had a sister who was a perfectly commonplace person, and believed everything she was told. Well! your sister would have been happily married and had a numerous progeny, and that would have been the survival of the fittest; but *you* would have been burnt in an auto-da-fé, and there would have been the end of you. You would have been unsuited to your environment. There! that's Evolution! Good-bye!" On went his hat, and we heard the hall-door close after him before we had done laughing."

The presence of such Victorian luminaries as Lyell, Browning, Tennyson, and Lord Shaftesbury at her dinner parties in the 1870s and 1880s gives some indication of the social circles in which Cobbe moved. It seemed that she knew everyone, and everyone knew her, and despite the "unsuitability" noted by Lyell, she was a respected and accepted member of the London establishment. Different, but not too different, she still shared enough of the values of her peers to remain one of them. Later in life, her bitterness and autocratic ways were to overwhelm her political savvy, but at this point Cobbe was firmly entrenched as the leader of a popular cause and a favorite hostess and guest.

In 1874, soon after the publication of Burdon Sanderson's *Handbook*, Cobbe created a petition addressed to the governing committee of the RSPCA to protest abuses of animals carried out in the name of science. This Memorial, as it was called, urged "immediate efforts for legal restriction of vivisection" and was signed by six hundred people, many of whom were in positions of great cultural influence. Signatories included clergymen (the Archbishop of York, Cardinal Manning), literary figures (Tennyson, Browning, and Carlyle), and "41 doctors, 24 Members of Parliament, 26 peers, 9 bishops, the Lord Chief Justice and the Mayor of London." The RSPCA, handed a powerful gift by Cobbe in the illustrious list, failed to act on the Memorial.

"Colam [the Secretary of the RSPCA] and the RSPCA refused to respond to Cobbe's initiative with more than investigation and a call for judicious inquiry," comments Richard French, author of *Antivivisection and Medical Science in Victorian Society*, the authoritative text on the subject. French argues that the RSPCA's caution on the vivisection issue, exemplified in its lukewarm response to Cobbe's Memorial, led to the founding soon after of a more focused antivivisection movement in England.

"The years 1875 and 1876 saw the beginnings of the hysteria and sensationalism in the antivivisectionist cause that were to force the RSPCA into an arm's-length relationship [with Cobbe and the antivivisectionists] and to discredit the movement in the eyes of a significant proportion of the press and the public. The society repeatedly dissociated itself from emotional accusations and called instead for carefully authenticated charges. Presumably its leadership was entirely satisfied to see separate antivivisectionist organizations developing, for extremist pressures might well provide leverage for concrete reforms through the good offices of the moderate majority of the animal protection movement."

In the autumn of 1875, while the Royal Commission on vivisection appointed by Queen Victoria gathered evidence and interviewed witnesses, the fifty-three-year-old Cobbe founded a society to work for the restriction of vivisection. Most of the founders of Cobbe's organization, soon known as the Victoria Street Society, were in agreement that to work for abolition was impractical; Cobbe herself appeared to share this view. However, when the restrictive bill introduced by Lord Carnarvon following the commission's deliberations was modified on the receipt of a petition signed by more than three thousand British physicians and scientists, Cobbe and her allies were outraged. In November 1876, Cobbe informed her colleagues that she would remain a member only if the group admitted a goal of outright abolition. On November 22, the Victoria Street Society adopted a resolution stating that "the society would watch the existing Act with a view to the enforcement of its restrictions and its extension to the total prohibition of painful experiments on animals."

Explaining this change of strategy in her autobiography, Cobbe wrote that before the adoption of the 1876 Cruelty to Animals Act, she and her supporters had "accepted blindly the representation of vivisection by its advocates as a rare resource of baffled surgeons and physicians, intent on some discovery for the immediate benefit of humanity or the solution of some pressing and important physiological problem; and we thought that with due and well-considered restrictions and safeguards on these occasional experiments we might effectually shut out cruelty. By slow, very slow degrees, we learned that nothing was much further from the truth."

After the scientists fought and successfully resisted Carnarvon's bill, which would have banned experiments on dogs, cats, and horses and legislated the use of anesthesia in all other experiments, Cobbe and her group decided that half measures would never be effective in preventing animal suffering because "vivisection we recognized at last to be a *method of research* which may be either sanctioned or prohibited as a method, but which cannot be restricted efficiently by rules founded on humane considerations wholly irrelevant to the scientific inquiry."

By insisting on abolition, Cobbe created dissension. A few of the society's supporters, recognizing that an absolute ban on animal experimentation was an unachievable goal, resigned, leaving the abolitionists in control. Nonetheless, Cobbe grew increasingly autocratic and intransigent, refusing all advice that deviated from her own designs. Her antagonism extended not only to her adversaries in science and Parliament, but also to those who shared her goals. She was particularly critical of Anna Kingsford.

The rivalry between Anna Kingsford and Frances Powers Cobbe weakened a movement that might have benefited from their combined efforts. John Vyvyan, an antivivisectionist author whose book *In Pity and in Anger* provides the fullest account, apart from Maitland's biography, of their dispute, notes that "if they had been able to cooperate, their gifts would have been complementary, and they might have done a spectacular work together. But this would have required from both of them a degree of self-understanding and self-discipline that neither possessed; and without that, they seemed predestined to enmity."

Cobbe was, as Vyvyan notes, a "bulky, practical, robust spinster" while Kingsford, twenty-four years her junior, was "beautiful, emotional, a wife and mother at the age of twenty-one, and with so delicate a constitution that she was almost constantly ill." Although both women possessed literary gifts, Kingsford's work tended toward the mystical and the poetic, while Cobbe had both feet firmly planted on the earth and wrote as a hardheaded (although not hard-hearted) social reformer.

Despite her apparent fragility, Kingsford aggressively pursued her chosen course. Often ill and alone, she spent six years in Paris studying medicine in a environment that was far from hospitable. Graduating in 1880, she presented her thesis, *"L'Alimentation vegetale de l'homme,"* in which she established, using the science of the day, that a vegetarian diet best promoted human health. This thesis was later expanded and published as a book, *The Perfect Way in Diet: A Treatise Advocating a Return to the Natural and Ancient Food of Our Race.* That Anna Kingsford was able to complete a six-year course of medical study in one of the capitals of animal experimentation without having vivisected a single animal and then persuade her professors to accept such a controversial thesis is testimony not only to her intelligence but also to her formidable will.

Increasingly, however, that will was directed toward ends that appeared freakish to many of her contemporaries. Antivivisection was a thoroughly respectable cause, as witnessed by the solidly bourgeois and aristocratic composition of Cobbe's coalition. But Anna Kingsford and Edward Maitland were drawn toward another group that inhabited the fringes of the Victorian establishment. Whether they defined themselves as spiritualists, Theosophists, or simple mystics, these were people who had moved far beyond traditional religious perspectives into realms of unconventional thought and practice. Kingsford was increasingly drawn to this group, and by 1883 she was president of the London branch of the Theosophical Society. Kingsford's penchant for occult activities like automatic writing, in which she served as a channel for spiritual insights granted by discarnate entities, her vegetarianism, and her willingness to speak openly about all of these things appalled Cobbe, who objected not only to Kingsford's mystical Christianity, but even more to the unconventional behavior it seemed to condone.

Although both Cobbe and Kingsford accepted the Victorian view that women were especially well-suited to campaign against the evils of vivisection by virtue of their inherent moral superiority, Cobbe embraced other elements of the traditional view that Kingsford—in her words and in her behavior—rejected. When Kingsford requested Cobbe's assistance in joining a London women's club and sponsorship in reentering London society after her extended sojourn abroad, Cobbe refused. Kingsford and Maitland were in Paris at the time, and in his biography of Kingsford, Maitland gives an account of Kingsford's dismay when she received Cobbe's response.

"Her amazement, therefore, was as great as her distress was keen when she received for answer an abrupt refusal to act in any way as her sponsor in London, on the ground that, having a family and home of her own, she had adopted a profession and a career which, in the writer's opinion, were incompatible with her domestic duties." Maitland urged Kingsford to write to her husband and enjoin him to express his support for her activities. Algernon Kingsford, apparently a rather remarkable character in his own right, then "wrote to Miss Cobbe a strong rebuke, denouncing her conduct as in the highest degree impertinent, cruel and wanton, inasmuch as every step taken by his wife had his full concurrence; and that in the event of any overt action to her detriment, he should deem it his duty to seek legal protection and redress."

Algernon Kingsford's threat of legal action notwithstanding, Anna Kingsford and Edward Maitland found themselves virtual pariahs in London society in a short time, "confronted by obstacles of a nature altogether unanticipated in the shape of a personal persecution of a most malignant kind," Maitland wrote heatedly in his biography of Kingsford. "Of the real motive, I was in no doubt, having sufficient insight into the character of the persecutrix to recognise her as capable of indulging any amount of jealousy of one whose endowments bid fair to make her a formidable rival in the cause with which Miss Cobbe had identified herself."

Although the opposition of Cobbe doubtless prevented Kingsford from achieving all that she would have wished, she was quite active in the antivivisection and Theosophical movements following the awarding of her medical degree in 1880. In 1882, she delivered an address before the British National Association of Spiritualists, called "Sorcery in Science," and published an article, "The Uselessness of Vivisection," in *The Nineteenth Century*, an influential journal. In 1883, she published a series of articles in *The National Reformer* and organized a Paris Antivivisection Society, with the author Victor Hugo as president. She and Maitland spent much of 1883 traveling in Switzerland, where Kingsford delivered lectures at Berne, Lausanne, Montreux, and Geneva. Her final publication was a letter to the editor of the *Pall Mall Gazette* denouncing the cruelty inherent in the production of fur and other fashions.

By then, Kingsford had met and befriended both Helena Blavatsky and Annie Besant, two prominent Theosophist writers and mediums. To Blavatsky she confided that she had long been working for the death of prominent vivisectors, with positive results in two cases. Maitland's account has been repeated by other antivivisectionist authors. In December 1877, while a student in Paris, Kingsford had a terrible argument with one of her professors, who, according to Maitland, "forced her into a controversy about vivisection, the immediate occasion being some experiments by Claude Bernard on animal heat, made by means of a stove invented by himself, so constructed as to allow of observations being made upon animals while being slowly baked to death." Experiments like these are described in Bernard's notes and published works, so there is no doubt that Kingsford's professor provided an accurate account of Bernard's activities.

On returning to her rooms that evening, Kingsford remained highly agitated, "and seeing in Claude Bernard the foremost living representative of the fell conspiracy, at once against the human and the divine, to destroy whom would be to rid the earth of one of its worst monsters," Kingsford "invoked the wrath of God upon him, at the same moment hurling her whole spiritual being at him with all her might, as if with intent then and there to smite him with destruction." On returning home that evening,

Maitland found her prostrate, and concludes in his book that "so completely, it seemed to her, had she gone out of herself in the effort that her physical system instantly collapsed, and she fell back powerless on her sofa, where she lay a while utterly exhausted and unable to move."

Two months later, Kingsford and Maitland, walking in Paris, passed the *Ecole de Médecine* and saw a notice posted announcing Bernard's death. For the rest of her life, Kingsford was convinced that she had killed him. In 1888, Bernard's onetime pupil Paul Bert died, and Kingsford congratulated herself on another death sentence carried out, because she had been actively working for his death, and that of Pasteur, since she had dispatched Claude Bernard. Until her own death at the age of forty-one, Kingsford continued to promote this policy of "spiritual assassination" and even suggested to Helena Blavatsky that "a band of occultists should combine in using their powers against vivisectors in order to destroy the system of vivisection, and ultimately those who persisted in practicing it." But Blavatsky declined, saying, "Let us work against the *principle* then; not against personalities. For it is a weed that requires more than seven, or seven times seven of us to extirpate."

By the time of Anna Kingsford's death in February 1888, the gospel of antivivisection had spread to the United States. As in England, the movement arose in response to the threat of massively increased research on animals, based on the needs of the new experimental sciences. Between 1870 and 1910, animal experimentation was to become an increasingly important aspect of medical education. The incorporation of laboratory training into the medical curriculum at Harvard in the 1870s and the founding of the Johns Hopkins School of Medicine in 1893 were key events in this process in the United States. The new sciences of bacteriology, immunology, and pharmacology demanded animal experimentation.

"Although presaged by the activity of men like Dalton and Flint, the opening of Henry Bowditch's laboratory at Harvard in 1871 marked the real beginning of experimental medicine in the

United States," Turner notes. "The succeeding steps came slowly. As late as 1880, only about a dozen scientists in perhaps five places carried out physiological investigations. . . . By the middle eighties, they were reinforced by some six or eight bacteriologists, who also experimented on animals. But only after 1890 did the United States become a major center of animal experimentation."

As in England, the first antivivisection societies were established by women, and women made up the bulk of the membership. Although Henry Bergh, who founded the American Society for the Prevention of Cruelty to Animals in 1865, was himself an antivivisectionist, after his death in 1888 the society retreated from his radicalism and adopted a more moderate stance on the issue. That left the fight to women like Caroline Earle White, who had helped found the Pennsylvania Society for the Prevention of Cruelty to Animals in 1867. White, the daughter of a Quaker abolitionist lawyer, met Frances Powers Cobbe on a trip to England and started the American Antivivisection Society on her return to the United States in 1883. According to medical historian Susan Lederer, "although membership in the society remained small compared to the much larger animal protection organizations, it commanded attention through the publication of the *Journal of Zoophily*, edited by White and published under the joint auspices of the PSPCA and the American Antivivisection Society. Perhaps more significant, Caroline White's assistant editor, Mary Frances Lovell, also served as the superintendent for the Department of Mercy for the Women's Christian Temperance Union, the largest women's group in the United States, which effectively spread the gospel of antivivisection to women's clubs throughout the nation."

As antivivisection sentiment spread throughout the United States, state and national societies were organized. Lederer charts the progression of organized efforts to combat the practice. The New England Antivivisection Society was founded in 1895, followed by the Antivivisection Society of Maryland in

1898, the Vivisection Reform Society in 1903, the Society for the Prevention of Abuse in Animal Experimentation in 1907, the California Antivivisection Society in 1908, and the Vivisection Investigation League in 1912. Although men often served as directors of these organizations, in keeping with the social norms of the time, women made up the bulk of the membership. Historians and sociologists have proposed a number of explanations for the importance of this issue for women of the time. Women were generally regarded as the guardians of morality and ethics and thus well suited to campaign against vice of all kinds. Frances Powers Cobbe, for example, "believed that women, although less intelligent and of course physically weaker than men, were more sympathetic, more direct and practical, less hypocritical, more facile in self-expression, and most importantly, more religious and moral than their masculine counterparts," says French.

Altruism alone (whether innate or socially constructed), however, seems insufficient to account for the passionate objections to vivisection felt by many women even if it explains why they were compelled to speak out on this issue in a culture that did not generally reward female assertiveness. Some writers have suggested that in a kind of political and psychological transference, the women who joined antivivisection societies to protest the suffering of animals and the abuses inflicted on them were expressing their own grievances, and that the antivivisection movement provided a safe, socially acceptable vehicle for their rage.

In *The Old Brown Dog: Women, Workers and Vivisection in Edwardian England*, historian Coral Lansbury posits that in the late nineteenth century, women's feelings of powerlessness, resentment, and anger over their subordinate status in society may have been sublimated into a profound empathy and identification with the suffering of experimental animals. In a provocative chapter, Lansbury draws parallels between violent sexual imagery common in Victorian and Edwardian pornography and the iconography of animal experimentation. In both cases, female human and animal "victims" were bound and helpless, in the former case subject to sexual torture, and in the latter, physiological investigation.

The Contagious Diseases Acts passed in England in 1864, 1866, and 1869 tend to support theorists who propose a link between women's fears about the medical establishment's attempts to control female sexuality and animal experimentation. The Acts established state regulation of prostitution, with compulsory medical examination and detention in "lock hospitals" for women infected by venereal disease. Men who patronized prostitutes were neither examined nor incarcerated, although they were equally likely to be infected. "Women alone suffered the degradation of internal inspection, and loss of liberty," sociologist Mary Ann Elston has noted, describing the view of women activists who agitated for the repeal of the Acts. Many of the same women were sympathetic to the antivivisection movement.

These parallels are also noted by French, who says that public campaigns against antivivisection, compulsory vaccination, and the Contagious Diseases Acts all "appealed to the same kinds of fears of, and hostilities toward, science and medicine." Practically speaking, "all three [movements] were campaigns against the increasing claims of science and medicine to the right to dictate morality and personal behavior." Certain medical procedures commonly performed on women patients of the era may have further established a link (whether conscious or unconscious) in the minds of many women. Oophorectomy or ovaritomy, removal of the ovaries, was recommended as a palliative for numerous female complaints, including "hysteria." "The death rates from ovaritomy were initially so high as to warrant the charge from within the profession, as well as from the outside, that it was an experiment for the benefit of the surgeon, not the woman," according to Elston.

Elston notes that near the end of the century "the metaphor of medical science, and medical practice on women, as rape, became a dominant theme in antivivisection literature, especially that written by women, from the 1880's onward. Women were explicitly invited to identify themselves with the animals, as potential victims of sexual assault by materialist medical men." In the United States, a similar theme was sounded by the feminist writer Charlotte Perkins Gilman, who wrote a fictionalized account of her treatment by the neurologist S. Weir Mitchell in "The Yellow Wallpaper." In Gilman's work, the protagonist, a young married woman diagnosed with hysteria, is forbidden by her doctor husband to read, write, or engage in any intellectual labor and is confined to her bed. As a result of this cure, the woman goes insane.

"Much of the treatment prescribed by physicians for hysteria reflects, in its draconian severity, their need to exert control—and when thwarted, their impulse to punish. Doctors frequently recommended suffocating hysterical women until their fits stopped, beating them across the face and body with wet towels, ridiculing and exposing them in front of family and friends, showering them with icy water," writes Carroll Smith-Rosenberg in "The Hysterical Woman: Sex Roles and Role Conflict in Nineteenth Century America." Mitchell's treatment of his "hysterical" women patients was no less severe, although the severity took another form. "At first, and in some cases for four or five weeks, I do not permit the patient to sit up, or to sew or write or read, or to use the hands in any active way except to clean the teeth," he wrote. "Where at first the most absolute rest is desirable, I arrange to have the bowels and water passed while lying down, and the patient is lifted onto a lounge for an hour in the morning and again at bedtime, and then lifted back again into the newly-made bed. In most cases of weakness, treated by rest, I insist on the patients being fed by the nurse, and when well enough to sit up in bed, I insist that the meats shall be cut up, so as to make it easier for the patient to feed herself."

This infantilizing of adult women who, according to Gilman, suffered from nothing more than the effects of institutionalized sexism and their anomalous position in a rapidly changing society reflects the greater struggle of women of the time to assert their authority and independence in a culture that granted them neither. "The hysterical female emerges from the essentially male medical literature of the nineteenth century as a 'child woman,' highly impressionable, highly labile, superficially sexual, exhibitionistic, given to dramatic body language and grand gestures, with strong dependency needs and decided ego weakness," Smith-Rosenberg notes. "But in a very literal sense these characteristics of the hysteric were merely hypertrophied versions of traits and behavior commonly reinforced in female children and adolescents. At a time when American society accepted egalitarian democracy and free will as transcendent social values, women, as we have seen, were nevertheless routinely socialized to fill a weak, dependent and severely limited role. They were sharply discouraged from expressing competition or mastery in such 'masculine' areas as physical skill, strength and courage, or in academic or commercial pursuits, while at the same time they were encouraged to be coquettish, entertaining, non-threatening and nurturant. Overt anger and violence were forbidden as unfeminine."

The antivivisection movement provided a seemingly heaven-sent opportunity for nineteenth-century women to flout male power and prerogatives at the same time that it enabled them to feel safely nurturant and feminine. Like other movements of the nineteenth and early twentieth centuries that both appealed to women's "superior" moral sensibilities and provided an escape from the narrow circle of domestic concerns (for example, temperance), antivivisection was both a womanly crusade against "male" vice and an opportunity to exercise "masculine" capabilities of leadership and public power. The movement gave these women a cause to fight for and a vehicle to express their own rage and resentment at their "caged" lives. Because they could not fight for themselves, they fought for animals, against the individuals and institutions they felt constituted a common oppressor.

—THREE—

THE DOGS OF WAR

It is a well-known fact that there are no industrial, no economic problems which are not related to problems of health. The better conditions of living, housing, working conditions in factories, pure food, a better supply of drinking water, all these great questions, social, industrial and economic, are bound up with the problems of public health. . . . It is of vital importance that health activities should be based upon accurate knowledge of the cause and spread of disease.

—William Henry Welch, M.D.,
address to the National Conference
of Charities, 1915

WASHINGTON, D.C., 1919

WILLIAM HENRY WELCH climbed the broad white steps of the Capitol with a martial step, his back straight and his senses sharpened for combat. The sixty-nine-year-old physician had arrived in Washington the night before to meet with his peers, the deans and professors of the nation's premiere medical schools, to plan their testimony before the Sixty-sixth Congress. On Saturday, November 1, 1919, the joy of the armistice still infused the nation's capital. Citizens of the city greeted each other with smiles and handshakes; they laughed and joked about everything, released at last from the tense vigilance of the preceding four years. The war against Germany had been won by the United States and its allies, but a key battle in the war against Welch's own sworn enemies was about to begin in the beautifully paneled Senate hearing room. He had no intention of losing this important engagement.

As he reached the top of the stairs and prepared to enter the building, a colleague stepped forward, cigar in hand.

"Ready to engage the enemy?" he asked.

Welch nodded and shook hands with Simon Flexner, his friend and former pupil, director of the Rockefeller Institute for Medical Research in New York City, founded in 1904. United by their commitment to the principles of scientific medicine and their belief that disease was a community problem, not just an individual calamity, the two men had often discussed the need for a new kind of physician-scientist, trained to study the spread of disease in populations. By the start of the new century, it was clear that infection spread fastest in the overcrowded and filthy tenements of the poor, where inadequate nutrition, lack of sanitation, and lax hygiene combined to create a breeding ground for pestilence. Welch and Flexner believed that a scientific approach to public health was the only way to prevent or arrest the spread of epidemics like the Spanish flu, which had killed 550,000 Americans in 1918, and the outbreaks of measles, pneumonia, and meningitis that had spread rapidly through Army training camps during the war. More common, if less massively lethal, epidemics of typhoid fever, tuberculosis, and malaria could be controlled. But public health officers in most cities in the United States, not trained in science, were ill equipped to combat disease and its causes. They were social workers and party hacks who too often received their positions as political favors and were completely unqualified for the work.

"How is faculty recruitment at the new institute coming along?" Flexner asked.

"Very well, very well," Welch replied cheerfully, and Flexner nodded, impressed as always by the vigor and enthusiasm of the man affectionately known to a generation of medical students as Popsy.

Welch's thirty-year-old dream of founding an institute to train public health professionals in the new sciences of statistics, nutrition, bacteriology, and physiology had finally materialized through the efforts of Simon Flexner's brother, Abraham. Abraham Flexner chaired the General Education Board of the Rockefeller Foundation, and when the foundation decided to award funds to create the first school of public health in the United States, he worked quietly to ensure that the money was awarded to Welch and Johns Hopkins University. The deans of Harvard and Columbia University howled, but Flexner prevailed, and in 1916 the Johns Hopkins School of Hygiene and Public Health was founded in Baltimore, with Welch, founding professor of pathology at the Johns Hopkins School of Medicine, as director. That's why he was here in Washington today, to counter once again this animal-worshiping nonsense that threatened to destroy the scientific approach to public health he had just been given the opportunity to create.

"No notes?" Flexner teased, as they entered the rotunda together, knowing that Popsy seldom prepared written notes even though he sometimes pretended to do so.

"No notes," Welch answered. "I'll be speaking from the heart today." Immediately, he felt vaguely embarrassed. He was a scientist, and as all of his colleagues knew, talk of the heart and emotions did not come easily to him. It must be the effect of the peace, he thought, and the sudden gush of relief that had loosened everyone's reserve.

"There they are," Flexner said as they entered the hearing room. "The Ladies' League Against Vivisection."

Welch glanced across the hearing room and saw his enemies, the soberly garbed members of Mrs. Farrell's animal protection coalition: society matrons, clerics, and professional busybodies of both sexes who had persuaded their allies in the Senate to introduce a bill prohibiting the vivisection of dogs in the District of Columbia and all territorial possessions of the United States. The bill would not affect Welch's operation at Johns Hopkins, since it did not address animal experimentation in the forty-eight states. But Welch and his colleagues knew that their adversaries would not be satisfied with a single victory. The D.C. dog bill was the wedge they would use to push through the ranks of scientific medicine. First dogs and then horses; first the District and then neighboring Virginia and Maryland. Like the Kaiser's army, they would continue to press forward unless defeated decisively on the field.

As Welch and Flexner took their seats, the chairman of the subcommittee banged his gavel to call the meeting to order. Following a brief introduction, he invited the sponsor of the bill, Henry L. Myers of Montana, to read his handiwork. The Senator's rich voiced filled the room.

"Whereas the dog has made a wonderful war record, and from everywhere word comes of his courage, his faithfulness, his cheery comradeship, and his keen intelligence; and whereas he has been decorated for bravery, serving his country, following its flag, and dying for its cause; Now as an act of right and justice to the dog and as a tribute to the soldiers who speak and plead for him, be it enacted by the Senate and House of Representatives of the United States of America in Congress assembled, that from and after the passage of this act it shall be a misdemeanor for any person to experiment or operate in any manner whatsoever upon any living dog, for any purpose other than healing or curing of said dog of physical ailments, in the District of Columbia or in any of the Territorial or Insular Possessions of the United States."

"Brilliant," Flexner whispered in Welch's ear.

Welch nodded, forced to agree. Although the scientists would never admit it publicly, their adversaries had conceived a strategy worthy of the military strategist Clausewitz. Defeated in previous attempts to enact congressional bans on animal experimentation in the early years of the century, the antivivisectionists had decided this time to limit their bill solely to the dog, an animal with a stellar war record. How, the antivivisectionists asked, could Congress permit experimentation on war heroes? Dogs had served the cause of freedom during the Great War. Were Americans now prepared to take away their freedom, to strap them to cold metal tables and torture them for science?

Welch focused his attention on the man on the stand, Walter

S. Hutchins, former president of the Washington Humane Society, who was introducing Edward Clement, president of the interstate coalition opposed to vivisection. Clement offered the opening remarks for his allies: "I must call attention to the fact that though it perhaps seems impertinent to introduce such a bill as this amid the great commotions and crises that the country faces today, on reflection one sees that these very commotions mean that the world is coming into a high state of spiritual exaltation, with sentiment coming into its own in all fields. Materialism and its prudences and calculations have been discredited, even in this businesslike country, ever since it took the high resolve that European civilization should not go down before the brutal arrogance of military superiority. If ever pure sentiment, pure idealism, operated to set in play the mighty forces of a great nation, it was when we entered the war against Germany without interest of our own."

Here it comes, Welch thought. He felt Flexner tense in his chair. Last night the scientists had discussed their greatest fear as they dined and talked strategy. If the antivivisectionists tapped into the poisonous well of anti-German sentiment in Congress and among the American people, the cause of science might well be placed in jeopardy. Germany may have been the enemy in the Great War, but as far as Welch, Flexner, and their peers were concerned, German scientists had been an elite army in the war against disease. Important research results flowed regularly from German laboratories, and Welch, like many of his peers, had learned most of the science he knew by studying with the great German teachers. He was basing his new school on a German model, the Institute of Hygiene in Munich, founded by Max von Pettenkofer. He held his breath and again focused on Clement.

"In other ages of the world torture has been regarded as a legitimate instrumentality for the extortion of evidence in the courts; and again, as an instrument of religion for the conversion of heretics, but in both fields it was outlawed centuries ago. Only in science is torture still tolerated and defiantly justified," Clement concluded. Welch and Flexner exchanged glances. Now they were on firmer ground. This slander could be countered. The dangerous moment had passed.

The next witness was Mrs. C. P. Farrell, representative of the sixteen animal protection societies petitioning Congress. She began with the usual assertion that she and the other antivivisectionists did not mean to attack the medical profession as a whole. "Each one of us knows splendid, humane, self-sacrificing men in the medical profession for whom we have the greatest respect," Mrs. Farrell said. "Many, indeed, are interested in the success of this bill. I am here to speak on vivisection and vivisectors, a body of men who, we are convinced, do not represent the medical profession as a whole."

Welch smiled. He had in his pocket a statement by the American Medical Association, the nation's largest professional body of physicians, testifying to the necessity of animal experimentation. This was going to be easier than he had hoped. Then Mrs. Farrell dropped her bomb. Looking directly at Welch and Flexner, and quivering with indignation, she forcefully presented the argument they had dreaded. "Led by Germany—and there is no doubt that Germany has led medical science, particularly in America, for many years—each European country has had its enthusiastic vivisectors, and, strange to say, these men seem to be the leaders in the vivisection world. I may mention Goltz, of Germany, of whom it has been said . . ."

Here she consulted her notes. "No advocate of unrestricted experimentation, so far as known, has ever dared to print the full details of the Goltz experiment. But I may present you with the following from a paper published by Dr. Max Buch. 'In order to test the relative sensibility of the sympathetic and vagus nerves, we have conducted experiments on numerous animals. No narcotics were used.'" Mrs. Farrell read the latter sentence slowly and with great emphasis. "'The vagus was laid bare. A great difference in the sensibility of the two nerves was evident. In the case of the crural nerve any movement of the pincers called forth expressions of pain, and, when, during the experiment, a thread was drawn underneath the nerve the animal began to scream piteously.' " Mrs. Farrell's voice quavered. Ladies in the audience dabbed at their eyes with lacy handkerchiefs. The members of the committee looked concerned. Welch sighed, glanced at Flexner, and shrugged. It was going to be a long day.

IF ONE MAN could be said to serve as a bridge between the old medicine and the new in the United States, that man would be William Henry Welch. Born in Norfolk, Connecticut, in 1850, Welch was delivered into a family of physicians—his father, grandfather, and various uncles all enjoyed flourishing practices in the town. Welch was expected to continue the family tradition, but at first he rebelled. Graduating from Yale in 1870, Welch planned to become a classical scholar and teacher of Greek. He was also at that time an orthodox Christian who delivered an address at his commencement exercises on the conflict between religion and science that concluded, "As long as the noble nature and the selfish nature are mingled in man, we shall be unable to reduce his actions to a scientific system."

Unable to find a position in his chosen field despite his academic success at Yale, Welch tried teaching and found it wanting before finally giving in and working as his father's apprentice for a brief period in 1872. He was miserable, however, and when his father proposed that he take a course in chemistry at the Sheffield School in New Haven, an orphan arm of Yale where scientific subjects were taught, Welch was thrilled at the opportunity to plunge once again into the stimulating intellectual atmosphere he had left behind. Later that year, after completing the course in New Haven, he entered the College of Physicians and Surgeons in New York City.

The medical education that Welch received at the College of Physicians and Surgeons was the best available at the time, but judged by the standards that Welch and his colleagues were to put into practice at Johns Hopkins twenty years later, both the curriculum and the method of instruction left much to be desired. There were no entrance requirements, and his teachers were community practitioners who imparted their impressions of various diseases and treatments in a series of lectures that students were expected to memorize. Students supplemented the lectures by practical experience gained as apprentices to practicing physicians.

In their biography of Welch, Simon Flexner and James Thomas Flexner noted that "of the required three years of study, only two had to be spent in medical school, and the terms each year lasted only six and a half months. The work was not graded, many pupils bought tickets for the entire course every year, and all attended the lectures in whatever order appealed to them. There was only one examination, the final one, which Welch found extremely easy. Indeed, the professors did not dare fail many of their pupils." This strategy made sense because if the students were unhappy with their teachers, for whatever reason, they could easily go elsewhere.

Even so, Welch was fortunate in coming into contact with a few good teachers, one of whom, Edward Seguin, had studied with the famed French neurologist Charcot. It was Seguin who gave Welch a microscope, even though it would be years before he learned to use it. Nonetheless he persevered, spending nights reading in the library of New York Hospital, which had the second most extensive collection of medical literature in the country, and studying pathological anatomy with Francis Delafield, who had himself studied with the famed French pathologist Louis. Under Delafield's tutelage, Welch spent long days and nights in "the dead house," performing hundreds of autopsies, which Delafield felt "were the source of most medical knowledge."

In 1876, Welch set off for Germany to study microscopic anatomy. It was an important trip that would not only determine the course of Welch's own life but also have a significant effect on the development of scientific medicine in America. In Germany, Welch found real laboratories staffed by what seemed an army of teachers, students, and assistants. Unlike the tedious recitation and regurgitation of facts common in American medical schools, students in Germany were expected to learn by doing, in the laboratory. In the basic course on histology that Welch took soon after arriving, each student was assigned a microscope and expected to pursue his own studies, with minimal guidance or assistance from the professor. "We have nothing in America like these laboratory courses. For example, in New York physiology is taught only by lectures, while here there is an excellent physiological laboratory where one can do all the experiments and study the subject practically," Welch wrote in a letter home.

Studying the subject of physiology practically, of course, meant observing the normal functions of the body via animal experimentation, and it was doubtless in Strasbourg, where Welch studied physiology and physiological chemistry, that he first became accustomed to the practice. He wrote to his sister in February and described his work with Professor Carl Ludwig, "my ideal of a scientific man, accepting nothing upon authority, but putting every scientific theory to the severest test." A few months later he wrote, "My work with Professor Ludwig has been very profitable, especially in giving me an insight into the apparatus and methods of modern physiology, which is by far the most exact of any of the branches of medicine, this position of exactness having been obtained more through the efforts of Prof. Ludwig than of any living man. He and the great Frenchman, Claude Bernard, are undoubtedly the two greatest living physiologists. Bernard is a more brilliant genius, but Ludwig surpasses him in exactness and really scientific investigations."

Later, in the laboratory of Julius Cohnheim, Welch learned to apply the methods of physiology to the study of pathology—to use animal experimentation to study changes produced in tissues and organs by disease. His first project involved oedema, or swelling, of the lungs, which often produced death by suffocation in human patients whose presenting symptoms differed widely. Although many medical writers had speculated on the causes of this phenomenon, the topic had not yet been experimentally investigated, until Welch tackled the job under the guidance of Cohnheim. "I will not weary you with the details of the experiments which I have made," Welch wrote to his father in 1877.

But the details he did provide made it clear that Welch was firmly convinced of the utility of animal experimentation and was using the methodology in his own studies. "I proved first that it is possible to produce oedema of the lungs in animals by obstructing the outflow of blood from the pulmonary veins, either by tying the aorta near the heart or by tying the veins of the lung. But . . . I found that it was necessary to make the obstruction perfectly enormous, so that for example the outflow of blood from the left ventricle must be almost obliterated before any change takes place in the blood pressure in the lungs which I measured with an instrument."

Later in the same project, Welch wondered how it was "possible to paralyze one side of the heart and not the other? We tried different heart poisons but as a rule they weakened both ventricles simultaneously. Finally, I hit upon a method which was as simple as the original idea. It was to open the chest [of a dog or rabbit] and expose the heart and to take the left ventricle between my fingers and squeeze it without injuring the right ventricle. To my delight it had the desired effect and the result was oedema of the lungs." Welch's investigations led him to conclude that oedema occurs in the last moments of life whenever the left ventricle of the heart stops functioning before the right, no matter what disease or condition originally acts to weaken the heart.

By the time Welch returned home to the United States in 1878, he was thoroughly convinced of the need to instill in American physicians the scientific knowledge and techniques he had learned in Germany. He had also been converted to Darwinism and the philosophical materialism embodied in that doctrine. "I am sorry for those whose faith in God can only stand or fall with the truth of falsity of Darwinism, for it seems to me no longer doubtful but that the theory of evolution will prevail. Those who are best able to judge it, men of science, accept it almost uniformly," he wrote to his sister. A few months later, in a letter to his stepmother, he expressed a new belief, which more than any other illustrated the transformation he had undergone in Germany. The young scholar of Greek who had once passionately

proclaimed his belief in free will was now unshakeably committed to determinism. "As no human power can change the
inevitable succession of events in nature, so we are made to be as
we are by education, circumstances which surround us, and the
evolution of causes over which we have no control," he said.

By the time Senator Norris, the subcommittee chair, announced a ninety-minute recess for lunch at 12:30, Mrs. Farrell
and her colleagues had presented letters from the governors
of Arizona, Oregon, Idaho, North Dakota, New Hampshire,
Wyoming, Arkansas, South Dakota, and Tennessee supporting
the bill prohibiting vivisection on dogs in the District of Columbia. The antivivisectionists had read accounts of historical experiments to prove that vivisection inflicted pain on animals. They
had recited a long list of writers, churchmen, legislators, and
medical men who had opposed the practice. All of this was irrelevant to the bill, in Welch's opinion and apparently that of the
committee, whose members pressed Mrs. Farrell for dates and
locations of the experiments she described. The most damaging
testimony had been offered by a young man who rented rooms
adjacent to the District of Columbia College of Medicine and
testified that he had heard piteous howls emanating from the
third floor of the building at all hours of the day and night. This
was evidence, the antivivisectionists asserted, that experimentation on unanesthetized dogs was taking place in 1919 in the District. A messenger had been dispatched to the dean of the school,
subpoenaing his testimony before the committee.

During the lunch recess, Welch sat alone in an empty office,
preparing his statement. In 1900, when he had last testified on
this issue before Congress, and delivered what he felt was a
crushing blow to his adversaries, he had been certain that the
issue had been laid to rest forever and that he would never be
forced to defend medical research in these halls again. The years
since he had returned from Germany had more than proved the
utility of animal experimentation, yet the penalty for violation of

the dog law was to be a fine of $100 to $500 per instance and/or imprisonment of three months to a year. Scientists and physicians could be jailed or impoverished for their work. Contemplating this injustice, Welch's outrage grew and his pen flew across the page as he prepared his testimony for the members of the committee. They were reasonable men. They must understand that dogs were not human, that the suffering of an animal, no matter how brave and beloved, could never equal the suffering of a family crushed by the loss of a child, unmoored by the death of a mother, or impoverished by the death of a father. Welch crafted his strategy and sent an army of words marching across the page.

"FOR THE SAKE of the record, please state your name, profession, business, and education."

"Dr. William H. Welch, Johns Hopkins University, director of the School of Hygiene and Public Health. Formerly dean of the medical faculty of the Johns Hopkins University, formerly Colonel, United States Army Medical Corps."

"Proceed."

Welch paused for a moment and looked out at the crowded committee room. Flexner nodded. Welch took a breath and began.

"I have been very much interested, Mr. Chairman, in this hearing. Twenty years ago an antivivisection bill was before the Senate, and at that time the matter was very thoroughly thrashed out. I think you will find extremely interesting things in the published report on the hearing in 1900 on Senator Gallinger's bill. I am particularly interested to see the change in strategy and the somewhat extreme attitude taken by the supporters of the bill presently before the committee. Twenty years ago no one would have admitted that they proposed the prohibition of animal experimentation. We knew that in their writings they did, but they were very guarded in any statement to that effect, that their real purpose was to abolish entirely experimentation on animals. And it is interesting to me that this purpose is frankly expressed here today, and that this bill, as Mr. Clement expressed it, is merely the thin edge, and they are hoping to secure abolition of experimentation upon animals."

Welch directed his gaze to one of the antivivisectionists who had testified earlier.

"I am also interested in the perfectly frank statement of Dr. Hutchinson that he is not aware of any benefit whatever that has been derived from experimentation on animals. Coming from a medical man, that interests me very much. It is also interesting to find the position taken here, frankly, that no matter how much benefit may be derived from the experimentation on animals and the relief of human misery, that it would not justify the use of animals for this purpose. It is a logical position, and I think it is good to have it presented, although I think the time is far distant when there will be any large conviction of public opinion to that effect. In testimony before the British Royal Commission on vivisection, there were certain advocates there of very restrictive measures, who said that if their children's lives depended upon the sacrifice of a guinea pig, they would not regard it as right that a guinea pig should be sacrificed to save the life of a child.

"This then," Welch continued in the same style he used to lecture to students at the university, "this is nothing new. The antivivisection agitation has been going on here longer than in Great Britain. It began in the sixties, and it has continued ever since. We, in this country, have done our best to safeguard this essential method of the advancement of medical knowledge, and up to this time no measure has been passed in Congress or in any state which would interfere with the legitimate use of animals for experimentation. If this bill is passed, it would be the first absolutely prohibitive measure."

Welch paused again, quickly looked down at his notes, and began his argument. "It is pathetic to have to comment upon some of the arguments presented today. Mrs. Farrell, of course, must try to make out a case that animal experimentation is of no benefit. I hesitate, almost, to take up those questions. I do not know that it is worthwhile. She is in opposition to the opinions of medical men and men of science. I admit that she has this little group of queer people that have served so long to be quoted from; but it is a hopeless case if they are going to argue from that point of view."

One of the members of the committee, Senator Colt, cleared his throat.

"Doctor, is it true that this experimentation is carried out in the methods described in the papers read today, subjecting the live dog to the most excruciating tortures?"

"No, Senator, it is not true," Welch answered forcefully. "We deny that entirely. I am willing to admit that there have been rarely unjustifiable experiments that I do not defend. But Mrs. Farrell goes back almost one hundred years, and antedates the discovery of anesthesia in her citations."

A woman in the back of the room stood and, before the chairman could react, shouted, "I have seen a dog's paw laid on a block, and I have seen it struck with a heavy mallet."

The room broke into disorder, with both the medical crowd and the antivivisectionists loudly asserting the truth or the falsity of this accusation. Senator Norris banged his gavel until all discussion ceased. He nodded at Welch.

"The witness may proceed."

Unperturbed, Welch took up the thread of his argument.

"I am perfectly willing to concede that there have been unjustifiable experiments, but we regard them as entirely exceptional. There is no use to which animals are put for the benefit of mankind, not food nor sport nor fashion, where there is equal care and solicitude to guard against the infliction of unnecessary pain as in experimentation."

"To what extent is the infliction of pain, if at all, necessary?" Senator Norris inquired.

"That is a difficult question to answer generally," Welch replied. "There are instances where the animal survives after it comes out from under the anesthetic. It is true that in some of those cases, so far as we can judge—we do not know much about pain to anyone except ourselves, to the human being—it is possible that there is, and probable that there is, suffering. That suffering is for a serious purpose—to benefit mankind. It just dwindles down really, to the question of whether man is justified in using the lower animals for his own good and at the same time inflicting suffering on them. The amount of suffering inflicted in these experiments as compared with that in slaughtering animals, and in the sexual mutilation of animals, in the hunting of animals for adornment of women's clothing and hats, and so forth, the amount of suffering in the laboratory is not comparable."

Again discussion erupted in the room and again the chairman banged his gavel, calling the assembly to order. "It is a question of a few hundreds of animals in the latter case as against thousands and millions of animals in the other case," Welch continued, speaking loudly against the commotion as the chairman continued to bang his gravel. "This particular use of animals for the benefit of mankind has been singled out for special reprobation. So far as the balance of humanitarian sentiment, Mr. Chairman, is concerned, it is with us. It is a narrow vision that does not see that it is inhuman to leave babies and mothers of mankind to the ravages of disease, and the immense suffering, which through advancement of knowledge, even if there be infliction of pain upon animals, need never be. The side of humanity is not with them; it is with us."

The room erupted into a frenzy of shouts and jeers. Again, Norris banged his gavel and called the room to order, threatening to expel anyone who continued to disrupt the hearing. In the silence that ensued, Senator Myers posed a question.

"I would like to ask you, if I may, are most of these experiments on animals conducted under the influence of anesthetics?"

"They all are," Welch replied. "Unless the purpose of the experiment would be frustrated by the use of anesthetics or unless the experiment is too trifling to require it. You would not give an anesthetic for a hypodermic injection or for a very slight incision. If it is going to cause any considerable amount of pain, an anesthetic is administered."

Welch prepared to continue, but Senator Myers interrupted.

"Do you claim, Dr. Welch, that experimentation is now, at this time, and continually, resulting in any substantial benefits and advancement of science?"

"Never more so than in the history of the world," Welch replied.

"What are some of the more recent instances of that?" the Senator inquired.

"One of the great benefits of experimentation was the discovery of the mode of prevention of yellow fever," Welch answered promptly. "That was by Dr. Walter Reed. It has saved countless lives. It will eradicate yellow fever."

"Was that based on experiments on animals?"

Welch paused and smiled.

"Actually, that was based on experiments on human beings."

Laughter erupted in the hearing room.

"I think our opponents have no objection to our experimenting on ourselves as much as we like," Welch continued. "Humans were our subjects because we knew of no animal at the time that was susceptible to the disease. But many problems concerning yellow fever, including protective vaccination, remain unsolved. These unsolved problems are now approaching solution by animal experiments conducted at the Rockefeller Institute for Medical Research and could not have been resolved otherwise."

Welch looked around the room at his assembled colleagues and continued. "In this room is Colonel Russell, who has saved thousands of lives and an incredible amount of sickness by establishing the methods of vaccination against typhoid fever, and that is based upon vivisectional work in its scientific principles. In this room is Dr. Flexner, who has reduced the mortality from cerebrospinal meningitis from 75 to 25 percent, based entirely on animal experimentation. All of this is disputed by these people."

Welch nodded toward the antivivisectionists and continued. "But it is accepted by the medical profession, and Mr. Senator, I think you would have no difficulty convincing yourself it is established beyond all doubt. I wonder sometimes at the attitude of mind of those who are so eager to demonstrate that some great discovery which promises to save lives is worthless. But for their cause they must do it. They must try to convince you that there is no value at all in this work. And that is one reason that I say that their attitude is not a humane attitude."

"It seems to me," interjected Senator Colt, "that the strongest defense of this is proof that medical science is advanced; and in the second place, to prove or show that the animal is rendered unconscious of pain, at the time of the experiments, by the administering of anesthetics."

In the back of the room a hand was raised and the speaker recognized. Reverend Dr. Smith, who had spoken for the anti-vivisectionists that morning, stood.

"Mr. Chairman, might I respectfully suggest that Dr. Welch is not really speaking to the point," Smith said in exasperation. "It is not in favor of prohibition that we are here today; we are not discussing that. We would run along different lines if we were here to discuss prohibition. We are talking about dogs and dogs only. We will grant him all the other animals to go and do what he will with, but only dogs. I do not believe I have ever heard it stated that dogs were necessary. I do not suppose he would maintain that dogs are necessary."

"I should maintain that dogs are absolutely necessary," Welch replied. "Much of our physiological knowledge has come from experiments on dogs, and for some experiments no other animal can be substituted. Where such substitution is possible, it is done. May I call attention to the work of Dr. Dandy on the brain? There is a disease called hydrocephalus, in infants, generally terminating in death. Dr. Dandy has made an exceedingly brilliant study, and dogs have been essential in that work in determining the mechanism by which this disease is produced, and how it can be relieved. It has attracted the attention of the whole world. Dr. Dandy's experiments and their application have already saved the lives of infants and will save many more from this common and terrible disease."

Senator Norris interrupted Welch.

"Do you mean to say that he can now cure hydrocephalus?"

Welch turned to the Senator and nodded. "Dr. Dandy said to me before I left Baltimore last evening, 'I wish you would tell them to come and talk to me. I have saved a good many little children's lives by experiments I made on dogs.'"

Welch then concluded. "There are so many others to speak of, Mr. Chairman, that I will not say any more at present but leave the rest to my colleagues."

"You may step down, Dr. Welch," the Senator replied.

Returning to Baltimore later that evening, Welch thought of Elmer McCollum's rats. McCollum was a professor of agricultural chemistry and nutrition at the University of Wisconsin whom Welch had recently recruited to head the Department of Chemical Hygiene at the new School of Hygiene and Public Health. While at the University of Wisconsin, McCollum had begun carrying out rather elegant nutritional studies on rats. Although the connection between human and rat physiology seemed obscure to many, the Wisconsin researchers believed that the same nutrients were probably crucial for both human and rat health.

McCollum had begun his studies using cows but soon decided to switch to rodents. "They are easy to feed and house, and they have a short life span," McCollum had explained when questioned about his choice of experimental subject. "I can do many more nutritional studies in a much shorter time." He attempted to use wild rats at first, caught in local barns, but found them too aggressive and averse to human contact to be of use. Next he bought twelve newborn albino rats from a local pet dealer and commenced his studies, arguing that "the most promising approach to the study of nutritional requirements of animals was through experimenting with small animals fed simplified diets composed of purified nutrients."

McCollum's gamble paid off when he and his colleagues discovered two vitamins, which they named A and B, that seemed important in maintaining animal health. When McCollum accepted his invitation to come to Baltimore, Welch had to arrange for the transportation of McCollum's colony of sixty experimental rats, descendants of the twelve animals McCollum had bought for six dollars from a Chicago pet dealer. Those rats were now housed in the basement of the School of Hygiene and Public Health, interbreeding with rats from the Wistar Institute in Philadelphia and other sources. McCollum was continuing his experiments in Baltimore, conducting controlled nutritional experiments using hundreds of different diets and measuring their effect on the rats' health and growth.

Rats were rather unpleasant creatures, Welch mused as the train pulled into Pennsylvania Station in Baltimore. People cared about dogs, but rats and mice evoked very little sympathy. It was doubtful that the Senate hearing room would ever be filled with people protesting experiments on rodents, Welch decided. As he stepped off the train, he saw a small swift shadow dart into a corner of the station. Under the circumstances, rats might well become the experimental animal of the future, he thought.

In a book published in 1926 called *Microbe Hunters*, Paul de Kruif, a bacteriologist turned author of popular science books, described the excitement and intellectual thrill of the early bacteriologists as they slowly made the discoveries that would change the way people thought about disease and death. No longer were physicians relatively helpless in the face of disease, dedicated solely to relieving the suffering of patients. An arsenal of biological products was being produced in laboratories throughout the world, proven victors in the war against infectious disease. Paul de Kruif was honest about the numbers of animals that were sacrificed by the microbe hunters in their quest; perhaps because he was thoroughly convinced of the utility of animal experimentation, he felt no need to defend it.

De Kruif's enthusiasm for biomedical science was as infectious as the diseases he described, and *Microbe Hunters* inspired generations of young men and women to join the battle against disease. By the time the book was published in 1926, it was clear that animals were absolutely essential in that fight. As William Henry Welch noted in his testimony before the Sixty-sixth Congress in 1919, the claims of the nineteenth-century antivivisectionists that animal experimentation was useless could no longer be accepted by any objective observer. As the old century waned and the new era dawned, a new victory was announced every few years as researchers isolated the microorganisms responsible for specific diseases and looked for the magic bullets that German researcher Paul Ehrlich had predicted would revolutionize medicine.

Behring and Kitasato's development of diphtheria antitoxin in 1890 was followed by Kitasato and Yersin's discovery of the plague bacillus in 1895. One year later, two different groups of researchers developed a vaccine for this scourge of the middle ages. In 1896, Kolle introduced a vaccine for cholera. In 1921, Calmette and Guerin developed a tuberculosis vaccine. Then, in 1927, Theiler and Smith isolated a virulent strain of yellow fever; eight years later they introduced a vaccine produced in tissue culture. In 1937, Hans Zinsser developed a vaccine for typhus, and by 1940, Harold Cox (building on the work of Zinsser, Fitzpatick, and Wei) had learned to culture the microorganism in the yolk sac of the chick embryo, enabling mass production of the typhus vaccine.

Animals were used by these researchers in a number of ways. In the early stages of the work, researchers injected animals with infectious material to determine whether or not the suspect microorganisms produced the disease in question. Once this fact was established, they could focus on growing pure cultures of the pathogen in animal tissue and "passaging" the material in tissue culture to reduce its virulence. Then whole animals were used once again to test the vaccines developed through this process. Why did the researchers use animals? Because the animals worked as models for the human victims of disease. If an infectious disease, or a physiological abnormality shown to result in disease (for example, diabetes) was present in the human population, it was likely that the same condition was naturally present or could be induced in at least one species of animal, possibly more.

The animal model could then be used by the scientist to elucidate the connection between the physiological abnormality and the disease in question and help develop a way to correct the problem. Alternatively, studies of disease specific to certain animal species may throw light on disease processes in human beings. As the Flexners note in their biography of Welch, "The problems of infectious diseases in animals have all the intrinsic interest of those in man, and for the pathologist they have the added value that they can be studied far more completely than can diseases in human beings. The whole biological history of a disease of animals stands ready to be unfolded, for in the instance of an inoculable, transmissible disease every stage from the very inception to the final termination in death or recovery can be explored."

Nonetheless, the method does have its limitations, and these were evident to Welch, who reminded his students not to confuse the model with the original. "When he found that experimenters were being led far astray, as in the belief that they actually reproduced in laboratory animals acute lobar pneumonia, he reminded them that 'acute lobar pneumonia as it occurs in human beings is a very definite and well-characterized affliction both anatomically and physically.'" Pointing out that animals are not naturally subject to the same condition as human beings, Welch informed his students that "the bare statement that the pneumonia produced experimentally is in all respects identical with acute lobar pneumonia in human beings must be received with caution." As Welch indicated by teaching and example, animal experiments (in vivo studies) are but one aspect of a broader research agenda. By insisting that medical schools be allied with both hospitals and universities, he and the other founders of scientific medicine in the United States illustrated the importance of combining laboratory work, whether in whole animals or cell and tissue culture, with close observation of disease in human patients. Each of these methodologies (including autopsy) provides important information and is an integral component in determining the causes, prevention, and cure of all forms of human disease.

After World War I, the rat and mouse, in particular, began to assume increasing importance in laboratory studies for a number of reasons. Small, easily housed and fed, with tremendous procreative capacities, these animals offered researchers almost unlimited opportunities to study the working of mammalian systems. "*Mus musculus*, the house mouse of North America and Europe, is the experimental animal, par excellence, of modern biomedical research," noted Herbert C. Morse in "The Laboratory Mouse—An Historical Perspective." "By virtue of its manifold variations, convenient size, fertility, short gestation period, ease of maintenance, variable susceptibility and resistance to different infections, and exemplification of many diseases that afflict mankind, it has found a major place in the laboratories of geneticists, developmental biologists, immunologists, cell biologists and oncologists." Although mice had been used for experimental purposes since the seventeenth century, most notably by Harvey, Priestley, and the French chemist Lavoisier, the explosive growth in mouse experimentation in the early years of the twentieth century was fed by the development of inbred strains for genetic research, after the rediscovery of the work of Gregor Mendel in 1900, and the development of a truly American school of biology, according to Morse, who studied the development of mouse strains for laboratory use.

For many of the same reasons, the rat also became a popular research subject. The founding of the Wistar Institute in Philadelphia played an integral role in this process as a group of researchers there, headed by Henry Donaldson, the institute's first scientific director, began to breed and develop strains of albino rats standardized for research. "Donaldson and his team of investigators at the Wistar began efforts in 1906 to standardize the albino rat. Initially, the main intent was to produce reliable strains for their studies of growth and development of the nervous system. In reality, the work directly gave the broad foundation for the use of the rat in nutrition, biochemistry, endocrinology, genetic, and behavioral research, and indirectly, in many other fields of investigation," according to J. R. Lindsey in *The Laboratory Rat.* The rat colony at the Wistar Institute, which was begun with four pairs of albino rats that Donaldson brought to Philadelphia from the University of Chicago in 1906, "served as the initial testing ground for developing satisfactory cages and ancillary equipment, diets, breeding practices, and facilities for rats," Lindsey says.

In 1911, the Wistar Institute began to supply rats to researchers and institutions throughout the world, as did the Jackson Laboratory in Bar Harbor, Maine, with mice during the 1930s. The hundreds of strains of rodents bred and sold by these two institutions, strains that differed not only in appearance but also in behavior and susceptibility to certain diseases and conditions, were developed in the early decades of the twentieth century for use in studies of mammalian genetics and hereditary factors in disease. "The 1920's saw the advent of the vast majority of inbred strains employed throughout the world today," Morse notes. "In spite of the financial hardships faced by scientists from 1920 to 1935, a large number of inherited variations appeared in the many inbred strains under development. Through analyses of these aberrant mice and their normal counterparts, it was increasingly apparent that the developmental biology of mice and men was strikingly similar. Mutations appearing on well-defined genetic backgrounds made it possible to ask detailed questions about the mechanisms of gene action in mammalian cells. From our current perspective, it is not unreasonable to suggest that the major contribution of the mouse to our knowledge of genetics and biology has been through analyses of single gene effects."

The increasing focus on genes and possible hereditary mechanisms of disease mirrored a similarly growing fascination with the causes of social pathology during the first decades of the century. Science was increasingly viewed as a potential source of solutions not only for the problems posed by infectious disease but for a broad range of social ills, many of which were believed to be heritable. The range of funding for the development of inbred strains of laboratory animals points to a widespread interest in the potential outcome of this research. The Jackson lab, for example, was "funded by moguls of the Detroit automobile industry," according to Morse, while MacDowell, who developed a strain of mouse highly susceptible to leukemia, "was supported by the social elite of New York City." Foreshadowing the ideologically driven genetics research of the next decade, MacDowell's wealthy patrons were not primarily interested in funding cancer research, Morse says. Instead, "they were determined that he should demonstrate that alcoholism in the lower classes was due to faulty genes, which they (of course) did not carry."

By the time William Henry Welch celebrated his eightieth birthday in 1930, sitting with President Herbert Hoover as an international radio hookup broadcast the president's congratulations around the world, the belief that science could cure the ancient ills of the human species had captured the public imagination. Men like Welch had convinced the public that the fears of the antivivisectionists and moralizers were unfounded, that science and medicine could accomplish miracles, and that the death of a certain number of animals was a necessary sacrifice in order to continue the war against disease—just as the death of a certain number of men had been necessary to protect democracy during the Great War. Moreover, American science had overcome its adversaries and competitors, and the United States was now the leading country for biomedical research. In that moment of triumph, Welch refused to acknowledge the birthday celebration and honors as a personal tribute. Instead, he noted that "by virtue of certain pioneering work and through over a half-century of service I stand here to represent an army of teachers, investigators, pupils, associates, and colleagues, whose work and contributions during this period have advanced the science and art of medicine and public health to the eminent position which they now hold in this country."

Four years later, in April 1934, Welch died in the hospital he had helped found. He died as he had lived, firmly convinced of the power of science and medicine to create a healthier, happier world. This progressive and optimistic philosophy, so much a part of the era in which Popsy Welch lived and worked, was already beginning to be swallowed by its shadow. Neither science, medicine, nor the word *progress* itself would ever seem so simple again.

FOUR

NAZI HEALING

Beginning with general hygiene and the fight against infectious disease, the development of public health leads from social medicine to racial hygiene. The medical profession has a special responsibility to work within the framework of the state on tasks posed by population politics and racial improvement.

—*Deutches Ärtzeblatt,*

June 1933

WIESBADEN, 1936

APRIL 20, 1936, was to be a great day for Germany. To honor Hitler's forty-seventh birthday, a congress was held in Wiesbaden to celebrate a new chapter in the illustrious history of German medicine. The congress was to inaugurate a new era, one in which academic scientists and physicians and Germany's traditional healers would work together to enhance the health of the nation. Over two thousand physicians and natural healers were meeting to bridge the gap between folk healing and biomedical research and create a New German Science of Healing that would deal with the whole person, not simply with diseased organs.

Prevention was to be the focus of the new science of healing. Individuals must assume responsibility for their own health. They must eat healthy foods, exercise, and swear off alcohol and tobacco. It was essential that these corruptions of modern industrial urban societies be banished. The healthy rural lifestyle of preindustrial societies was the model for the new Germany.

Mechanistic Jewish medicine and decadent internationalist lifestyles were ruining the health of the people, just as decadent Expressionist painting and theater had ruined German art, the Nazi propagandists said. The soul sickness of modern society could only be healed by a return to roots, to the traditional values and practices of the Aryan people. Midwives, herbalists, and homeopaths, with their deep roots in German folk medicine, had to be recognized as legitimate healers. Eventually they would be licensed by the state and would practice with Aryan physicians. It was the responsibility of every German to be a healthy cell in the body of the *Volk*.

Vivisection was now banned in Prussia and Bavaria and severely limited elsewhere. Like kosher slaughter, vivisection was seen as an example of Jewish cruelty to animals, a creation of the mechanistic Jewish mind, which saw nature as something to be dominated and torn apart. Germans worshiped nature as a beautiful whole and sought to understand it organically, holistically. The bond between humans and animals, who were part of the living soul of the land, was said to be sacred to all true Germans. On August 17, 1933, Hermann Göring announced that persons found vivisecting animals of any kind would immediately be transported to concentration camps. There was some grumbling among academic physicians about this ban, but it was squelched, just as those who protested that the natural healers were charlatans and quacks were asked to reconsider their views. They were assured that charlatans would be weeded out in the licensing process and that only the legitimate healers would remain. Occasional animal experiments might still be carried out, but they were strictly regulated.

Organic living was the new ideal. The Reich Research Council was investigating the use of pesticides in Germany. Some wanted to ban DDT, fearing that it was having an adverse effect on the genes of the German people. Other scientists were researching toxins in food, cosmetics, and in the environment. These products of decadent modern civilization were also believed to cause cancer and damage the gene pool. Hitler himself had promised plentiful funding for this research. The lasting health of the *Volk* was considered far more important than the ephemeral pleasures and passions of individuals. Individuals would die, but the Nation would live to triumph. The new Germany would be 100 percent organic.

THE WORD *eugenics*, formed from two Greek roots, *eu* (good) and *gen* (birth), was coined by the nineteenth-century statistician Francis Galton to refer to selective breeding within human populations. Galton, sometimes called the father of eugenics, was cousin to Charles Darwin and was influenced by Darwin's theory of natural selection. Galton believed that if scientific principles were rationally applied to populations, a healthier, more harmonious society could be created. He argued that humankind could take charge of its evolution and that this could be achieved by encouraging the wealthy, the successful, the beautiful, and the intelligent, collectively termed the fit, to reproduce. Galton and his followers believed that nations should discourage or impede childbearing by the destitute, the physically weak, the mentally ill, and others deemed unfit.

This philosophy appealed to the upper classes of Galton's day, who attributed their privileged lives to superior biological endowments. Social Darwinism, with its belief that culture is a mirror of nature and that certain individuals, groups, and races achieve and maintain cultural and economic hegemony due to their superior fitness, meshed well with Calvinist religious doctrine, which equated success with virtue. Galton's ideas appealed to social reformers on both the left and the right and had a broad following in England, Europe, and the United States by the turn of the century. Eugenics societies were founded on both continents, with socially prominent professionals of all political persuasions advocating eugenic policies. In the United States, President Theodore Roosevelt, birth-control pioneer Margaret Sanger, and inventor Alexander Graham Bell were all eugenicists.

By the second decade of the twentieth century, eugenics and

"racial hygiene" were academic disciplines taught in prestigious universities and discussed in scholarly journals on the continent, in England, and in the United States. Eugenics was a common topic in popular magazines as well. Scholars studied "race," attempting to group and categorize human beings with scientific precision and to trace the genetic basis of social problems such as mental illness, poverty, crime, and alcoholism. The research was interdisciplinary—anthropologists, statisticians, biomedical researchers, and physicians all pursued studies and published papers on the topic. The focus of the biomedical research was inherited illness, and its goal was to define and eradicate disorders believed to be genetically based.

The belief that evolution could be directed by human beings, and that "unfit" individuals and groups should be eliminated to improve the human stock, reached its apotheosis in Nazi Germany, but it was by no means confined to that nation. In the United States before World War II, twenty-five states passed laws permitting the involuntary sterilization of the mentally ill, criminally insane, and retarded. Congress passed strict immigration laws in 1921 and 1924 to limit entry of southern and eastern European immigrants—Italians, Poles, and Russian Jews—who were believed to be intellectually and morally inferior to Americans of northern European descent. Miscegenation laws forbidding marriages between blacks and whites passed in the United States in the same period are another manifestation of the same philosophy, similar to the bans on Jewish-German relationships in Nazi Germany. The intent of both the German and American laws was to prevent genetic contamination, the mixing of supposedly "superior" and "inferior" races.

This racist legislation was founded on a belief in "progressive" evolution, which supported eugenicist thinking. However, as Sir Charles Lyell pointed out to Frances Powers Cobbe's confused dinner guests, survival of the fittest does not necessarily imply survival of the best. Natural selection "proposes no perfecting principles, no guarantee of general improvement; in short, no reason for general approbation in a political climate favoring innate progress in nature," according to a contemporary defender of Darwin's, the Harvard zoologist Stephen Jay Gould. Although Darwin's theory was used to buttress many utopian pipe dreams regarding the perfectability of humankind, in its pure form the theory of evolution runs absolutely counter to any such philosophical or biological Pollyannaism. In the 1930s, however, the progressive fallacy was widespread; postwar condemnations of the "Nazification" of German science and medicine in the 1930s often ignored the pervasive belief in eugenics and its influence on policy in the rest of Europe and the United States in the twenties and thirties. While it is true that the German government went much further than any other nation in actually carrying out eugenic "cleansing," other nations also dabbled in eugenic legislation and certainly were home to large numbers of citizens who subscribed to the philosophy.

By 1939, the Nazi Hereditary Health Courts had reviewed close to 400,000 proposals for the sterilization of German citizens. The majority of these petitions were granted and the sterilizations carried out. A bureaucracy was established to review and implement the sterilization program. Eighty-one genetic health courts were set up to handle the business of receiving and processing proposals. The courts were composed of three members, two of whom were physicians and the third a district judge. Vasectomy and tubal ligation were the preferred methods; however, removal of the uterus was approved in some cases, as was experimental X-ray sterilization, which often resulted in severe burns and death.

In 1938–39, eugenic sterilization in Germany was succeeded by euthanasia. The first victims were children. The test case was a baby born physically and mentally handicapped, blind, with one leg and part of an arm missing. The child's parents requested a mercy killing. The petition was approved and authorized by Hitler, with instructions that similar cases could be handled in the same manner. Within a year, all children under three suffering from a variety of handicaps, including hydrocephalus, malformed limbs, and paralysis, were being executed, with or without parental consent. In October 1939, euthanasia was extended to older children and handicapped and mentally ill German adults.

Questionnaires were distributed to all German psychiatric hospitals, clinics, and institutions housing the mentally or chronically ill. Physicians in these institutions were required to fill out forms on every patient. After a cursory "review," patients targeted for euthanasia were transferred to one of the six killing centers set up throughout the country. Some patients were killed by lethal injection, but most were killed by carbon monoxide poisoning in the first Nazi gas chambers, complete with fake shower rooms. Physicians selected the patients and physicians turned on the gas. Approximately sixty thousand persons were killed in the euthanasia program.

Autopsies were often performed on the victims. Scientists and physicians not actively involved in the euthanasia program frequently benefited indirectly by receiving samples of tissues and organs. Professor J. Hallervorden, a prominent neuropathologist, received many brains from the killing centers. "There was wonderful material among these brains, beautiful mental defectives," he said later. "I accepted these brains of course. Where they came from and how they came to me was really none of my business."

Testifying at the Nuremburg trials after the war, the physician Karl Brandt, who supervised the euthanasia program, said, "It is possible as has been here frequently claimed, that euthanasia has something inhuman about it. But this is merely an outward appearance and lies mostly in the application. Euthanasia was there to help the sufferer alleviate the pain. These considerations can hardly be called inhuman, and personally, I do not find this unethical or amoral. . . . What I did, I felt was justifiable, and was supported by the most human concern. My only interest was to shorten the life, the painful existence of these miserable creatures."

In an article published in the *British Medical Journal* in December 1996, Hartmut M. Hanauske-Abel, assistant professor of pediatrics at Cornell University, traced the "convergence of political, scientific and economic forces" that led to the appalling abuses of science and medicine in Nazi Germany. The full details of that terrible story have only recently begun to emerge in the work of scholars within Germany and without. As Hanauske-Abel noted, the prevailing postwar view has been that these events had a kind of historical specificity. "The use of the term Nazi implies historical uniqueness, designates a chapter closed and finalised, and intimates that those involved had been 'Nazis' who, disguised as doctors, gained control and then executed their ideology." But as Hanauske-Abel and other scholars have noted, this was not the case, and the most important lessons of the Nazi years with respect to science and medicine have been virtually ignored, with potentially disastrous results. "The notion that something will not happen again prepares the ground for cataclysmic re-enactments."

The work of scholars such as Hanauske-Abel, Robert J. Lifton, Anne Harrington, Robert J. Proctor, and others reveals that certain themes that resonated with the medical and scientific establishment throughout the 1930s were not adopted by these communities as a matter of political expediency. Instead, they were a natural outgrowth of the science of the day. The most significant of these themes, a focus on the genetic basis of both physical and mental illness and a preoccupation with "wholesome" lifestyles, eerily foreshadow contemporary concerns. The National Socialist government officially sought to reduce the consumption of both alcohol and tobacco. Its research program focused heavily on the environmental effects of chemicals. Prevention of illness through healthy "natural" living was promoted as the ideal; indeed, according to the Nazis, German citizens had a responsibility to the nation to be healthy.

These policies sound reasonable. They grew out of long-standing traditions in German culture and science and expressed equally long-standing tensions. As Robert Procter, the scholar whose work has most carefully probed these roots of Nazi scientific policy and practice, has noted, "On the one hand, the Nazis wanted to return to what they saw as the original, natural state of human life and society. On the other hand, Nazi medical authorities also wanted to breed a better human, and this induced them to entertain radical measures to alter and 'improve' the course of human biological history."

In *Racial Hygiene: Medicine Under the Nazis*, Procter describes the Organic Vision of Nazi Germany and its connection to animal protection and environmentalism. This environmental and medical holism had a strong anti-Semitic element. "Nazi encouragement for new and more organic forms of medicine must be seen as part of a broader movement to restore more natural ways of living to the German people," Procter comments. Nazi physicians linked their concerns for the protection of nature and nature's creatures with the supposed need for a retreat from the impersonal, "capitalist," "socialist," or "Jewish" forms of medicine that had long dominated the earlier liberal age. Jews were blamed for having suppressed more natural German healing practices and replaced them with "internationalist medicine," Procter writes.

The ban on vivisection (never fully implemented, according to most scholars) was one element of this worldview. Not only was vivisection viewed as an aspect of a mechanistic scientific model that ignored the whole person in favor of treating diseased parts (identified as a "Jewish" approach by the Nazis), but Judaism itself was blamed for an ethic that sanctioned animal abuse. In *Regarding Animals*, sociologists Arnold Arluke and Clinton R. Sanders devote a chapter to National Socialist attitudes toward animals and animal protection legislation. The essential unity of humans and nature, a recurrent theme in German culture, provided a basis for the stringent animal protection legislation passed by the Nazis in November 1933. These laws "permitted experiments on animals in some circumstances, but subject to a set of eight conditions and only with the explicit permission of the Ministry of the Interior, supported by the recommendation of local authorities," Arluke and Sanders note. "The conditions were designed to eliminate pain and prevent unnecessary experiments. Horses, dogs, cats, and apes were singled out for special protection. Permission to experiment on animals was not given to individuals but only to institutions." Laws against vivisection and kosher slaughter were followed by wildlife legislation in 1934 and 1935.

Arluke and Sanders conclude that the Nazis' identification with and cult worship of animals, together with animalization of humans and elevation of the moral status of nonhuman animals "helped to shape the Third Reich's criticisms of scientific thinking and such practices as vivisection." This blurring of the boundaries between human and animal led to both an increased empathy and respect for certain species of animals and a willingness to adopt the "ethics" of the animal world. If humans were a part of nature, and not distinct from it, and the Jews with their excessively analytical system of morality and science were the primary culprits in turning humankind away from what was "natural," then of course the Jews, and all forms of unnatural "Jewish" thought, must be purged.

These authors, together with Robert Procter and other scholars, agree that Hitler and many of his chief aides were vegetarians who believed that abstaining from meat would not only enhance human health but also spiritually regenerate the human race. Procter, Arluke, Sanders, and others have attributed Nazi dietary habits to the influence of Richard Wagner, the nineteenth-century German composer, who believed that civilization could be regenerated through vegetarianism. Procter states unequivocally in *Racial Hygiene: Medicine Under the Nazis* that "Hitler was a vegetarian and did not smoke or drink." In *Regarding Animals*, Arluke and Sanders note that "many leading Nazis practiced vegetarianism. Hitler hired a vegetarian cook and became very critical of others who were not vegetarian, sometimes referring to beef broth as 'corpse tea' and sausages as 'cadavers.' " Arluke and Sanders provide several quotes from Goebbels's diary, including "Meat-eating is a perversion of our human nature. When we reach a higher level of civilization, we shall doubtless overcome it."

Many contemporary vegetarians vehemently dispute these scholars' claims. Animal activist Rynn Berry, who serves as historical adviser to the North American Vegetarian Society, said in a 1999 interview that he was motivated to explore this issue after he published a book called *Famous Vegetarians: Antiquity to Present* and found that "people were constantly twitting me about why Hitler was not included."

Berry, who recently published a pamphlet called "Why Hitler Was Not a Vegetarian," believes that Hitler's vegetarianism was a marketing scheme concocted by Nazi propagandists who wished to create a public perception of Hitler as a kind of "fascistic Gandhi, a self-denying ascetic forgoing animal flesh for the good of the nation." Although Hitler "paid lip service to vegetarianism, he never practiced it." Scholars who promote the view that Hitler was a vegetarian tend not to be vegetarians themselves, Berry says, and tend to define vegetarianism inaccurately. Although Hitler's doctors put him on a meat-free diet in 1938 because of failing health, Berry says that people close to Hitler recalled after his death that the Führer's favorite meal included liver dumplings.

"Hitler was in no way an ethical vegetarian," Berry asserts. He believes that it is important to counter the assertions of scholars that the chief Nazi abstained from meat "because nonvegetarians tend to use the Nazi issue to discredit vegetarianism in general." Like many people who have sworn off meat, Berry is convinced that doing so confers spiritual as well as material benefits. "The Buddha said that eating animal flesh extinguishes the seeds of compassion," he notes, dismissing the notion of Hitler's vegetarianism as a "myth." The unwillingness of contemporary animal activists and vegetarians like Berry to believe that the most sadistic and brutal regime of modern times was run by a group of mystical vegetarians who believed, like Anna Kingsford, that the human race could be redeemed by a vegetarian diet is shared by less ideologically minded people who cannot reconcile the Nazis' remarkably humane attitude toward animals, exemplified in negative attitudes toward hunting, meat eating, and vivisection, with their vicious treatment of human beings.

This paradox poses an important question that remains unanswered: Does an elevation in the moral status of animals inevitably result in a degradation in the moral status of human beings? Certainly in Nazi Germany such appears to be the case. Once the distinction between human and animal had been leveled, it became possible to conceive of breeding humans in the same way that a conscientious farmer might breed his animals, to improve the health of the stock. Diseased or weak or otherwise unwanted humans could be disposed of with the same lack of moral feeling that one might experience in "putting down" sick or surplus beasts. Certain species of animals could be valued over certain "races" of human beings, depending on their utility or aesthetic or symbolic value. If human life and animal life were morally equivalent, acts that had previously been unthinkable could become commonplace. One would not hesitate to destroy a kitten born blind and legless. So why permit a human baby with the same deficits to live? Old, sick animals were useless on a farm. They sapped resources—just like the chronically ill and handicapped people in institutions. Best not to waste precious food and other resources on nonproductive, therefore superfluous, bodies. From there it was only a short step to viewing certain types of human beings as literal vermin—dirty, disease-spreading parasites—that could be exterminated with as little remorse as one might feel in crushing a sewer rat.

THE NAZI FASCINATION with race and national character was entwined with an obsession with public health. In the 1920s, infectious disease remained a threat. But the disease that inspired a special dread in the years immediately before World War I and afterward, providing both a metaphor and a motive for Nazi persecution of the Jews, was typhus. Following Charles Nicoll's 1909 announcement that typhus was spread by lice, and Howard Taylor Ricketts's discovery the next year of the pathogen that infected the lice and was then passed on to their human hosts, certain "outsider" groups began to be stigmatized and associated with the dirtiness that bred disease. This was true in the United States, where boatloads of new arrivals from eastern and southern Europe poured into the cities of the Northeast, settling into crowded slums, and in Europe, where Jews remained an outsider population, although long established.

Historians of science and medicine believe that nearly thirty million people were infected with typhus during World War I, and that over three million people died of the disease. The bacteriologist Hans Zinsser estimated in his book *Rats, Lice and History*, published in 1934, that nearly three million died of typhus in Russia alone from 1916 to the end of the civil war. In Germany the dread of typhus developed into a near phobia, with "Jews, lice, and typhus" a Nazi slogan expressing the belief that Jews were dirty, disease-ridden creatures who would infect the German people.

"Out of respect for human life I would remove a purulent appendix from a diseased body. The Jews are the purulent appendix in the body of Europe," commented Fritz Klein, an Auschwitz doctor, echoing Nazi ideology, which equated Jews with infection, disease, and death. Nonetheless, these human "bacilli," as Hitler termed the Jews, proved useful as slaves for German industry—and as experimental subjects. When the German army was menaced by typhus soon after the start of the Russian campaign, experiments on new vaccines were begun in January 1942 at Buchenwald concentration camp. One group of prisoners was injected with experimental vaccines while a control group received no vaccine. Both groups were then infected with typhus—the same types of experiments traditional in vaccine research, but using human rather than animal subjects.

A range of experiments were conducted at various camps: sterilization studies at Auschwitz-Birkenau, in which victims genitals were irradiated or removed; high-altitude studies at Dachau, where prisoners were subject to pressure so extreme that eardrums burst and organs ruptured; hydrothermia experiments, also at Dachau, in which prisoners were immersed in freezing water or tied naked to boards covered with only a sheet and repeatedly doused with freezing water before being left outside during winter; gangrene and sulfonamide experiments at Ravensbrück, in which women's legs were cut to the bone and infected with glass and wood splinters. The catalogue of procedures is endless, limited only by the imaginations of the physicians and scientists who conducted them. Bodies of prisoners were treated as scientific "material."

When questioned about their activities after the war, German physicians who participated in these experiments seldom expressed guilt or remorse for their activities. For the most part, they claimed to be following orders or working in the service of science or the state. As Fritz Ernest Fischer, who participated in the experiments at Ravensbrück, commented at his trial, "If one desires to judge a particular kind of behavior, one has to consider the aims which underlie it. I hoped by means of these experiments to obtain certain results by which many wounded people might be helped . . . I committed these acts as an obedient member of the military forces. The law, the Führer, and the government, so I perceived it then, ensured me of legal protection. My personal responsibility played no role in this at all."

Psychoanalyst Robert J. Lifton and political scientist James M. Glass have attempted to trace the development of Nazi policy on euthanasia and sterilization to its culmination in the Final Solution, the genocide of the Jewish people. In *Life Unworthy of Life: Racial Phobia and Mass Murder in Hitler's Germany*, Glass charges that "it was not cultural propagandists who organized the infamous 'special treatment' of the Jews; it was the public health officials, the scientific journals, the physicians, the administrators, and the lawyers, who feared that the very presence of the Jews would endanger their families, their bodies, and ultimately their lives. To think of the Jew in such terms is insane from our perspective, but it was held to be sane in a culture caught up in the phobic projection of infection onto the Jews and the scientific authority legitimizing such beliefs."

Jews were viewed as carriers of sickness, Glass says, a self-fulfilling prophecy when millions were herded into ghettos and forced labor camps in Poland and overcrowding, hunger, and unsanitary conditions created the necessary conditions for epidemics of infectious disease. "It was not that the Germans were unaware of typhus as a disease process; their approach to treatment and containment indicate a knowledge of its connection with public health. But the German mind-set did not link unsanitary living conditions ruthlessly imposed on the Jewish population with disease. Jews contracted typhus and sickness not because of being forced into intolerable physical environments— very few physicians admitted to this fact—but because Jewish blood and genes predisposed Jewish bodies to infection." Ghettos and labor camps were viewed by the German authorities as "places for isolating vermin," Glass says, and the reason they were set up outside the borders of Germany was to avoid contamination of German soil. "The Final Solution, from the perpetrators' perspective, did not *look* psychotic, but rather the opposite: it appeared a sensible, defensible precaution taken against a real public health menace."

The critical document that set the stage for both the murder of "unfit" Germans and the destruction of the Jewish people is identified by both Glass and Lifton as *Die Freigabe der Vernichtung lebensunwerten Lebens* or *The Permission to Destroy Life Unworthy of Life*, published in 1920 by two German professors, Karl Binding and Alfred Hoche. Binding, a retired law professor, and Hoche, a professor of psychiatry, identified certain groups as economic and social burdens. Lifton writes: "The book included as 'unworthy life' not only the incurably ill but large segments of the mentally ill, the feebleminded, and retarded and deformed children. More than that, the authors professionalized and medicalized the entire concept. And they stressed the *therapeutic* goal of that concept: destroying life unworthy of life is 'purely a healing treatment' and a 'healing work.'"

The efforts of Glass, Lifton, and other scholars working to uncover the roots of genocide in Nazi Germany raise troubling questions about the manner in which the ideals of science and medicine can be used to justify the most heinous crimes. Moreover, as both Glass and Lifton emphasize, the ideology of National Socialism was itself biologically based. "One can speak of the Nazi state as a 'biocracy,' " writes Lifton in *The Nazi Doctors*. "The model here is a theocracy, a system of rule by the priests of a sacred order under the claim of divine prerogative. In the case of Nazi biocracy, the divine prerogative was that of cure through purification and revitalization of the Aryan race. . . . The Nazi biocracy differed from a classical theocracy in that the biological priests did not actually rule. The clear rulers were Adolf Hitler and his circle, not biological theorists and certainly not the doctors. . . . In any case, Nazi ruling authority was maintained in the name of the higher biological principle." Nearly every scholar writing about the role of science and medicine in Nazi Germany quotes Rudolf Hess's famous dictum. "National Socialism is nothing more than applied biology."

In Nazi Germany, as Robert Glass puts it, "to care about public health meant caring about the health of the nation's blood. The public health professions defined care as a therapeutic cleansing of pollutants to Aryan blood and eliminating genetic embarrassments to the race. 'Burying differences' came to be understood as burying bodies."

ON NOVEMBER 20, 1945, an international military tribunal convened in Nuremburg, site of Nazism's great rallies. Although Adolf Hitler, Heinrich Himmler, and Joseph Goebbels were dead by their own hands, twenty-four other high-ranking Nazi officials, including Rudolf Hess, Hermann Göring, and Albert Speer, were brought to trial. Nineteen of the defendants were found guilty and twelve were executed. Seven of those found guilty, including Hess and Speer, received long prison sentences. From October 25, 1946, to August 1947, a second tribunal was held, in which twenty-three German scientists and physicians were charged with experimentation on human subjects in Germany and throughout the occupied territories. Fifteen defendants in this trial were found guilty; seven were executed and eight imprisoned.

The number of individuals prosecuted at Nuremburg fell far

short of the number of German citizens who willingly participated in the crimes of the Nazi regime. German physicians and scientists, in particular, were not only active supporters of Nazi philosophy and practice but intellectual architects of the sterilization and euthanasia programs, which led inexorably to the Final Solution. The great majority of those who reviewed documents, performed examinations, and signed papers that resulted in the sterilization and death of thousands of handicapped German adults and children, and the mass murder of millions of Jews, were never identified or prosecuted. The majority of German physicians who were members of the SS faded back into society after the war, setting up medical practices and otherwise returning to "normal" life. Nor were those whose intellectual contributions to Nazism (including the rhetoric of racial hygiene and eugenics) created the climate in which these events could occur ever brought to justice.

As James M. Glass has written: "What distinguishes the final Solution from other historical instances of mass murder lies in the mobilization of the scientific and professional sectors of the German population in pursuit of a culturewide phobia. It was a scientifically designated process begun long before actual combat, and, with the exception of the SS and concentration camp guards, sane people ordering and carrying out the atrocities. In fact, it was the physicians, administrators, and scientists at some psychological distance from the victims who created the ideology and values of the regime. No other genocide had its major impetus in the language of science, law, and administration of the society."

The horrific acts conducted in the name of racial hygiene, genetics, and research by scientists and physicians during the Nazi years led to the drafting of the Nuremburg Code, a ten-point statement defining the moral and ethical responsibilities of scientists seeking to conduct research on human subjects. Signatories agreed that subjects must be adult volunteers fully capable of understanding and consenting to experimental procedures. Despite the clear language of the code, the postwar period has been marked by debate about the exact nature of informed consent, particularly as it applies to vulnerable populations like military personnel, minorities, children, and the mentally ill, and punctuated by scandals in which the failure of researchers to obtain informed consent has been revealed to an outraged public.

The most infamous chapter in U.S. research history runs parallel to the German experience, illustrating both the pervasiveness of eugenic thinking in the prewar period and the willingness of some scientists and physicians to ignore ethical imperatives when they conflict with research goals. In 1932, the year before Adolf Hitler came to power in Germany, the United States Public Health Service began a study of the effects of untreated syphilis in black men. Three hundred and ninety-nine men in Macon County, Alabama, were unwitting subjects in this experiment, which was not suspended until 1972, when reporter Jean Heller of the Associated Press published an account of the study. Although at the time the study was initiated there was no real treatment for the disease and no legislative mandate to consider the rights of human subjects, both a cure for syphilis and an acknowledgment of the need to obtain informed consent were developed in the immediate postwar period. Nonetheless, the medical staff and public health officials running the study decided that the possible research benefits of continuing the study outweighed their moral obligations to offer appropriate treatment to the men, as described by James Jones in *Bad Blood: The Tuskegee Syphilis Experiment.*

Like the animals that were by then an integral part of a growing research industry, the men of the Tuskegee Syphilis Study were viewed primarily as a means to a scientific end. Racism certainly played a part in the tragedy, as did bureaucratic inertia, but the context in which the drama was enacted was the continued lack of oversight or accountability by biomedical researchers to anyone outside the closed community of science. Increasingly, it appeared that the only people who could understand science were scientists, and the authority of science was such that few outsiders were willing to argue for a voice in what were perceived by most as matters far beyond their understanding.

The German experience has been commonly viewed as an aberration, a bizarre chapter of history in which an entire nation went mad. Few were willing to see (and even fewer to say) that the seeds of that madness lay dormant in every nation. Some Americans knew the details of the atrocities, but they were not heavily publicized, and when a German psychoanalyst, Alexander Mitscherlisch, published a book in Germany that detailed the crimes of physicians and scientists, it was quickly suppressed. Flush with victory and prosperity, the United States prepared to march triumphantly forward into "the American century." These were to be golden years for American science, when the American people, secure in the sense of their own freedom and righteousness, tacitly approved whatever researchers did in their laboratories in return for the many benefits scientists promised—and by and large delivered.

FIVE

POLIO POLITICS

Animal experimentation in our country is assuming the proportions of a major industry whose very survival will depend upon sound humanitarian supervision. . . . I can only hope that you will try to put yourself in the place of one of the millions of experimental animals that we force to suffer for our comforts.

—Robert Gesell, M.D.,
January 1952

BOSTON, 1952

CHRISTINE STEVENS and her father, Dr. Robert Gesell, prepared to enter the research laboratories of a local medical school. They were expected, had submitted a petition to visit the animal quarters over six months ago, and had been instructed to appear at this specific date and hour. Still, even with this advance notice, they had been kept waiting for more than three hours. Gesell and Stevens were here as representatives of the Animal Welfare Institute, an organization they had founded in 1951 to promote the welfare of animals and to reduce the amount of pain and fear inflicted on animals by human beings. A physiologist, Gesell was particularly interested in improving the treatment of laboratory animals. He was dismayed by the conditions in which research animals were housed, confined to small, wire-bottomed cages in which they could neither turn nor stretch comfortably, enduring a cramped existence without exercise or companionship—and as a scientist he suspected that the experimental data gathered from such animals would be flawed at best. He knew that the health of experimental animals in some laboratories was often seriously compromised by infectious disease and parasites and that not enough attention was paid to such crucial factors as diet, exercise, and sanitation. Often, little was known about the animal's background and genetic history. Even if one were to take the most extreme perspective and look upon the creatures as nothing more than machines, laboratory equipment, the flaws in this system were evident. How could reliable results be gathered from subjects whose health status was poor and whose genetic variability virtually guaranteed a baffling range of biological responses to any experimental protocol?

"Would you like some coffee while you are waiting?" a secretary asked Stevens.

"No, thank you. We're fine," she replied graciously.

Stevens and her father exchanged a glance. They knew from experience that there was no point in protesting the long wait. Best to remain calm and pleasant. Their reputation as fair and unemotional observers was their ticket into the labs.

Nonetheless, the conditions they had observed in laboratories throughout the country disturbed them. As early as 1946, Gesell had written to A. J. Carlson, a fellow physiologist who had helped form the new National Society for Medical Research (NSMR), "I am not one of those who believe that the conditions of animal experimentation are ideal. I believe the commission could raise the question whether the experimental animal is receiving the consideration to which he is entitled, particularly as regards survival experiments in which the animal is likely to suffer. It is my experience that there are always a number of us who may be too sure of man's privileges to experiment on the lower forms. Some system of scrutinizing the soundness of biological problems and the skill and wisdom and consideration of the scientist would do much to convince the public that our minds are open to all sides of the problem. I doubt the wisdom of the policy which offers no supervision of animal experimentation whatever."

Gesell had hoped then that the NSMR would develop a system of oversight similar to the British system, but administered by scientists themselves rather than by the government. However, it soon became clear to Gesell, Christine Stevens, and her husband, Roger, that the NSMR had no intention of wading into the cloudy waters of animal welfare. The official stance of the organization was that anyone raising questions about the welfare of animals used in experimentation was a closet antivivisectionist. Disappointed by the resistance of the scientific establishment, Christine and Roger Stevens founded their own group, the Animal Welfare Institute, with Gesell and a number of other physiologists on its Scientific Advisory Board.

Writing again to Carlson that year to announce AWI's founding, Gesell compared certain Canadian and American animal experiments recently denounced in an editorial in the British medical journal *Lancet* to the acts of scientists in Hitler's Germany. "These ominous experiments make us search our souls and wonder what the future has in store for us for they remind us so inescapably of the Doctors of Infamy who performed terminal experiments on men and women without the use of anesthesia. These experiments become all the more ominous when it is taken into consideration that they were performed on creatures as sentient as man and in a free and happy country, and not under duress of a harsh government barbarized by global warfare."

For that reason, the newly founded Animal Welfare Institute, although interested in every aspect of animal welfare, "will, at present, concern itself particularly with humane safeguards in the use of animals for research and medicine," Gesell wrote. "On the urgent problem of laboratory animals, the Institute will have the following immediate aims: the establishment of a code for the handling and use of laboratory animals, the encouragement of medical research of the highest possible quality as opposed to mere quantity, and a survey of existing and proposed programs for the procurement of laboratory animals in an effort to discover the best possible method of satisfying the needs of humanity in the advancement of medicine, public health etc. without jeopardizing practical animal welfare work and the general advancement of humanity through increasing consideration for all living creatures."

It seemed a reasonable agenda to Christine Stevens, president of the organization, her husband, Roger, its treasurer, and to the members of the Scientific Advisory Board. However, the NSMR had accused Gesell, Stevens, and the members of the board of a concealed antivivisectionist bias. Paradoxically, the antivivisection organizations were equally critical of AWI's work, claiming that the institute was nothing more than a front organization for research interests, mouthing animal welfare concerns but secretly devoted to easing the road for experiments. The situation would be very discouraging, indeed, if it weren't for the letters Gesell and Stevens had received from physicians and scientists throughout the United States and abroad expressing quiet support for their endeavor. "That scientists should not be hampered and animals completely protected: this is one of those consummations devoutly to be hoped for," an American biologist had recently written.

Christine Stevens sat in the offices of the laboratory director and gazed out the window at the sunlight playing on the bright green grass. It was a mild spring day, and the light was perfect for painting. Stevens was an artist, but since she had founded the institute, she was spending less and less time at her easel. Instead she wrote the AWI's newsletter, handled its correspondence, testified in congressional committees, visited laboratories, and used all of her wit and intelligence to fight the NSMR's latest campaign, the passage of pound seizure acts throughout the country. The AWI had just lost the battle in New York, which had recently become the fifth state to empower laboratories to seize unclaimed cats and dogs and allocate them to laboratories.

The Metcalf-Hatch Act, signed into law by Governor Thomas E. Dewey on March 10, now permitted the Commissioner of Health to seize all unclaimed cats and dogs from any animal shelter receiving public funds and allocate them to the 475 laboratories in the state currently licensed to use experimental animals. Fifteen pounds operated by humane societies fell under the jurisdiction of the Act and had fought hard against its passage. But the NSMR and its local ally, the New York State Society for Medical Research, had fought harder, circulating petitions among hospital patients in support of the bill and encouraging physicians, medical and pharmacy students, and researchers to write to legislators urging its passage. In the midst of the controversy, the New York State Society had held a press conference to announce that the shortage of experimental animals was impeding research on the effects of atomic radiation at the Atomic Energy Commission's laboratory in Rochester. Fueled by public fears of the atom bomb and by the state society's sophisticated lobbying campaign, the bill passed. The only course open to Gesell, Stevens, and their supporters was to visit laboratories and monitor conditions, publishing the results of their investigations and attempting to improve the feeding, housing, and care of the confined animals. This they were not permitted to do in New York State, as the bill's supporters had successfully countered a proposal for regular laboratory inspections by outside observers.

"We're ready for you now," one of the animal handlers said. Stevens noted that, like many U.S. laboratory workers, he spoke with a heavy accent. Unlike their counterparts in Great Britain, which had in 1951 established a professional association that provided training for laboratory technicians, the people responsible for caring for animals in the United States were often viewed by scientists as low-level janitorial staff, responsible for feeding the animals and cleaning their cages but little else. "What we need is people to take care of the animals, not just to take care of the cages," an American biologist had ruefully remarked to Stevens on one of her recent visits. Often the lab workers were recent immigrants who spoke little English. Stevens made a note to herself as she and her father entered the lab and were confronted by the powerful odor of animal and recently applied disinfectant. The manual for the care of laboratory animals that she was currently writing must be published in both English and Spanish.

Polio Politics

THE POLK SCHOOL smelled. It smelled like an institution and it smelled like sickness. It smelled like exactly what it was, a state home for retarded adults and children. As Jonas Salk and his executive secretary, Lorraine Friedman, walked into the school on the morning of May 23 after driving 80 miles from Pittsburgh, laden with syringes, vials, index cards, and lollipops, the smell engulfed them. But they were not deterred by the smell or by the apprehensions of the immunization committee of the National Foundation for Infantile Paralysis, which had cautiously approved this trip. Dr. Salk had faith in his work, and Lorraine Friedman had faith in Dr. Salk. The director of the Polk School, Dr. Gale Walker, had also expressed confidence in the young researcher. In response to Salk's inquiry two years before, Walker had volunteered the children of the Polk School to be the first human beings aside from Salk, his family, and laboratory workers to receive the killed virus vaccine that might save millions of adults and children from the twentieth-century scourge: polio.

Walker greeted Salk and Friedman cordially, escorting them up the stairs to a bright sunny room with a desk and chair for Friedman. She was to enter the name of each subject and other important information on index cards and to assist Salk as he drew blood and then administered an injection of vaccine to each child. Then she was to give each child a lollipop. The lollipops were important.

"We're ready," Salk said. He and Friedman exchanged a glance. Salk was aware that some scientists would condemn this trial, but he had confidence in his vaccine and he was thoroughly familiar with vaccine trials, having received his training in the trial use of influenza vaccines among military personnel in the 1940s. It was time to see if his polio vaccine worked, if his injections could save Americans and their children from the horror of another epidemic in which hundreds would die and thousands would be left paralyzed.

Summer after summer beginning in 1916, parents and children throughout the United States had lived in fear. That first epidemic of acute infantile poliomyelitis (soon shortened by headline writers to "polio"), in which 6,000 had died and over 27,000 others had been left permanently crippled, had been followed by other summers of terror in which parents attempted to keep children closeted indoors, refusing to let them swim in public pools or drink from water fountains. Public health officials closed camps, schools, and movies theaters, and normal activities were suspended as terrified parents tried to keep their children safe from this invisible menace. In at least one case, a whole city (Annapolis, Maryland) was quarantined. In those first decades of the twentieth century, no one was really sure where this new disease had come from, nor how it was spread—only that it struck mostly children, who woke one morning with a fever and a stiff neck, and hours or days later found themselves paralyzed, often for life. Wards full of children lay in "iron lungs," whole-body respirators that forced air in and out of their damaged lungs.

But the disease did not strike only children. Franklin Delano Roosevelt had been a vigorous thirty-nine-year-old husband, father, and politician summering with his family on Campobello Island, Maine, in August 1921 when he was struck down. He already knew that he wanted to be president, but the disease left him permanently paralyzed from the waist down. Roosevelt's misfortune, which he turned into a personal and political asset, led to Salk's visit to the Polk School thirty-one years later. A few years after his illness, in 1924, Roosevelt had used a large portion of his inheritance to buy Warm Springs, a resort in Georgia popular with polio survivors who thought that bathing in the mineral-rich waters there might help them regain the use of their limbs. Roosevelt had grand ideas for the dilapidated resort, planning to turn it into a vacation mecca for the wealthy and famous, whose paid stays would help finance the medical care of the polio survivors who summered there. However, few Americans proved willing to vacation at a resort known to be a magnet for handicapped people.

Warm Springs was a financial disaster, and Roosevelt, absorbed by his political ambitions, asked his law partner Basil O'Connor to take over its management in 1926. O'Connor created the nonprofit Warm Springs Foundation in 1927 and launched a series of fund-raising efforts based on Roosevelt's increasing popularity. On September 23, 1937, the Warm Springs Foundation, flush with funds from a series of President's Birthday Balls to raise funds for polio patients and their families, became the National Foundation for Infantile Paralysis, the most successful private funding source for biomedical research in United States history. According to J. R. Paul, M.D., whose epic *History of Poliomyelitis* was published in 1971, "The initial impact which the NFIP made upon that minuscule number of established workers in the field of poliomyelitis . . . amounted to the sudden appearance of a fairy godmother of quite mammoth proportions who thrived on publicity."

Throughout the late thirties, the forties, and the early fifties, the NFIP raised millions of dollars for research into the causes and treatment of polio. Americans were urged to donate whatever they could to the battle against polio, and donate they did—sending dimes to the White House in such numbers that the halls of the president's home were blocked by bags of coins sent in the first March of Dimes campaigns. Driven by the fear that their children would be struck by the disease and urged on by the powerful public relations machinery of the NFIP, which kept that fear alive and used it to raise funds for research and the care of polio survivors, Americans poured millions of dollars into the coffers of the National Foundation for Infantile Paralysis. "Between 1938 and 1962, the Foundation's annual income was something like $25 million, or roughly $630 million over the whole period." Six years after its founding, NFIP director Basil O'Connor boasted that the foundation "has made 298 grants to 74 institutions involving 114 groups of workers, in one of the greatest scientific attacks against any disease."

By 1954, when Jonas Salk, a foundation grantee buoyed by positive results in experimental trials at the Polk School and the

D. T. Watson Home for Crippled Children, persuaded the members of the NFIP's Vaccine Advisory Committee that his vaccine was ready for a large-scale field trial, Americans volunteered their children as human guinea pigs in the largest scientific experiment ever conducted. The years of drumbeating by the foundation and the increasing respect of the American people for science and scientists made such voluntary large-scale human experimentation possible. As Paul comments, "One of the achievements brought about by the effective use of propaganda techniques was the creation of public interest in poliomyelitis research as a holy quest. The image of the disease as an evil thing that must be conquered and banished forevermore took hold." In later years, the vocabulary of war and conquest would be applied to other diseases, notably cancer and AIDS. But in 1952, the enemy was polio, and the nation had mobilized to fight it to the death. It was the birth of big biological science, and Jonas Salk was about to become a national hero.

NEW YORK, 1955

"THE CORNER DRUG STORE may have a good supply of the Salk polio vaccine in time for the 1955 polio season—thanks to 'heated competition' among commercial producers of the vaccine," enthused an article published in the *Pittsburgh Sun-Telegraph* on December 12, 1954. "Basil O'Connor, president of the National Foundation for Infantile Paralysis, told the *Sun-Telegraph* that the secrecy with which commercial drug houses are 'gearing' for the polio vaccine is a good sign. The race won't get underway, however, until the verdict on the vaccine's big field test last spring is given April 1, O'Connor explained."

This excerpt was reprinted in the January 1955 newsletter of the Animal Welfare Institute *Information Report*. Like everyone else in the United States, the staff and advisory board of AWI rejoiced at the prospect of a safe and effective polio vaccine. But they were not pleased, indeed they were horrified, by the transit conditions of the monkeys being imported for polio research. AWI staff and humane-society personnel inspected shipments of monkeys arriving from India and found overwhelming suffering, sickness, and death among the animals. Five airlines were transporting monkeys to the United States, with a single flight bringing as many as 2,200 recently captured animals into the country. The trade was highly profitable for animal dealers. The inspection groups found much to report.

"A shipment of 900 rhesus monkeys for a large pharmaceutical house from a dealer in Calcutta. Crates were wooden, 17"×18"×35", with wire top and front. Some contained as many as twenty monkeys, most had eighteen. These figures were stamped on the front showing the contents, but in some cases there were fewer in the cages, presumably because they had died en route and been removed. Because of crowded conditions which prevented us from seeing into the back of many crates, we could not make an accurate count, but we saw eight dead monkeys and many sick. . . . An official of the airline stated that between forty and fifty monkeys had died en route. In addition to the atrocious overcrowding described above (twenty monkeys in a crate 17 inches high with floor space 18 inchcs by 35 inches, means less than four by eight inches of floor space for each monkey), rough handling caused further suffering."

The report described the arrival of a shipment of 1,300 monkeys ordered from another dealer for another pharmaceutical house. "Walked long way to the plane. Found large van in front of it with steep steel chute connecting the opening in the plane to back of van. As we arrived we saw a crate start down and heard a tremendous crash at the bottom. Our escort from the airline company called out. When we came into the light he introduced us as 'ladies from the humane society.'" In the presence of the humane-society representatives, the employees of the animal dealer cleaned the cages. "Two men picked up each crate and tipped it so that the two-inch opening on the bottom turned toward the floor. They then gave the crate three or four hard knocks against the floor. This threw the monkeys all about and against the wire on the front of the crate. One monkey had lost the end of its tail, which was bleeding. Another monkey was almost dead in a crate, lying limply with diarrhea. Excrement dripped on the white bread which was given the monkeys on arrival. Monkeys reached for this bread with great eagerness, fighting over crumbs which fell into their crates before they got their share, but after a few bites they dropped it aside."

The AWI's scientific advisory committee added that "in addition to the cruelty these animals are compelled to endure, their diets are changed three or four times in their shift from freedom to captivity, which certainly contributes to weakening their resistance to the diseases that frequently decimate laboratory monkey colonies." The article noted that "cruelty and the economic waste from the resulting sickness and death of the monkeys go hand in hand."

The AWI reported on these conditions to the Bureau of Customs in New York and pointed out that they violated both United States law (Public Law 72, which provided for criminal penalties for anyone importing animals or birds into the country under inhumane or unhealthful conditions) and customs regulations demanding immediate investigation of inhumane transport conditions. The Bureau of Customs responded positively, inviting AWI to submit recommendations on proper shipping and suggestions for minimum standards. The AWI called on Frederick A. Ulmer, Jr., curator of mammals at the Philadelphia Zoological Gardens, to help draft its report on humane care of monkeys in transit. The government of India placed an embargo on the export of monkeys on March 10, 1955, pending further investigation. The embargo was lifted on April 5, less than a month later, when the National Foundation for Infantile Paralysis pledged itself to ensuring the monkeys' health and well-being during the flight to the United States.

The National Society for Medical Research was not pleased by the Animal Welfare Institute's interference in the business of monkey transport. The *Bulletin of the American Institute for Biological Sciences* of June 1956 included a statement by Ralph

Rohweder, executive secretary of the NSMR, comparing the staff of the AWI to Machiavelli, Hitler, and Stalin. "The practice of Machiavelli's rule of deliberate hypocrisy has been perfected over the years. Among the great artists of our time have been Adolph Hitler and Joseph Stalin. But using a beneficent mask for a malicious purpose is not limited to big-time politics. It occurs even in the relatively intimate organizational affairs of science. A representative example is the history of the Animal Welfare Institute."

Charging that the AWI had "carefully avoided the crazy headdress of the antivivisection cult" while "making all the points in the AV propaganda bag," Rohweder commented that "the AWI misses no opportunity to talk vividly of suffering and to make the utterly ridiculous claim that scientists somehow are exempt from the cruelty-to-animals laws." Rohweder charged that "agitation by the AWI helped bring about the embargo on the shipment of monkeys that seriously hampered the production and testing of polio vaccine last spring."

Christine Stevens reprinted Rohweder's editorial in her newsletter and took the opportunity to clarify AWI's position on the polio monkey situation. "Everyone hopes to see an end to polio, but if it is indeed the monkeys that bring an end to this disease, is it too much to ask that they be treated at least as well as individuals being sent to zoos or to individual private owners? The Animal Welfare Institute believes that animals which are to be sacrificed for human benefit deserve more, not less, consideration than those which, like the pets and animals for zoos mentioned above, are expected to live out their natural life span. It also believes that a part of the money so generously given by the American public to prevent human suffering from polio might very properly be expended in bringing about a reduction of the vast amount of wholly unnecessary suffering now being inflicted on monkeys used for polio research and vaccine production."

THE SEARCH for a polio vaccine was significant in a number of important ways. For one thing, it illustrates the manner in which biological science had become, by the early 1950s, an enterprise that demanded enormous sums of money for laboratories, supplies, and animals. The days when a scientist could equip an entire laboratory for $3,000, as William Henry Welch did when he created a pathology lab at Johns Hopkins, were long gone. The National Foundation for Infantile Paralysis, which encouraged Salk (an influenza researcher) to focus on polio, provided funds to construct Salk's laboratory and offices at the University of Pittsburgh and to employ a large staff of technicians, junior researchers, and an administrative assistant. The foundation began funding Salk in 1947. His original grants were to be used to discover how many types of poliovirus existed—work carried out in monkeys, then, as now, the most expensive experimental animal. Although monkeys, apes, and other nonhuman primates had been used in research before, the real boom in monkey studies occurred during the polio years. In *Of Mice, Models and Men: A Critical Evaluation of Animal Research*, Andrew Rowan estimates that "between 1953 and 1960, well over a million monkeys died, many en route to the laboratories, as Salk, Sabin and others raced to be the first to develop an effective vaccine."

Disturbed by the large numbers of animals that died in transit or arrived at research laboratories sick or so weak that they were vulnerable to numerous infections that spread rapidly in laboratories, the foundation set up a monkey "conditioning center" in South Carolina. Monkeys purchased for foundation research were sent to the center, called Okatie Farms, to recover from the stress of capture and transport. A few weeks or months later, these animals, well fed and healthy, would be sent on their way, less likely to carry and transmit diseases that could decimate an entire laboratory.

Studies conducted by Salk and his team and other researchers working on polio—Enders at Children's Hospital in Boston; Bodian, Morgan, and Howe at Johns Hopkins; Horstmann at Yale; Cox and Koprowski at Lederle Laboratories; Sabin at the University of Cincinnati—required an enormous number of monkeys. Early estimates by the National Foundation for Infantile Paralysis indicated that at least fifty thousand monkeys would be needed to carry out virus typing studies alone, before any vaccine research. Years later Albert Sabin, who created an oral polio vaccine licensed in 1961, said in a paper published in the *Journal of the American Medical Association* that approximately "9,000 monkeys, 150 chimpanzees, and 133 human volunteers" had been used by his laboratory in polio studies. "These studies were necessary to solve many problems before an oral polio vaccine could become a reality."

However, by the time Salk and other researchers had completed the typing of the virus in the early fifties and begun to work on a vaccine, another methodology had entered the picture, one that was not only to facilitate typing but also enabled researchers to produce the enormous amounts of virus necessary for vaccine development. This method was tissue culture— removing organs and tissue from animals (or humans) and keeping the material alive and growing for experimental purposes in culture. Although primitive in vitro methods had been used to create vaccines in the late nineteenth century, and individual scientists had been attempting to refine the process of tissue culture since then, it wasn't until 1947 that a conference in Hershey, Pennsylvania, led to the founding of the American Tissue Culture Association, which would systematically promote the development of the method. In that same year, John Enders, Thomas H. Weller, and Frederick C. Robbins grew mumps virus in tissue culture using penicillin to prevent bacterial contamination. The next year they made the discovery for which they would be awarded the Nobel Prize, showing that poliovirus could be grown in cultures of normal human embryonic cells. A few years later, in 1954, Dulbecco and Vogt successfully cultivated poliovirus in cell cultures of monkey kidney. Once the virus had been successfully propagated in culture, the preparation of a vaccine against the disease became possible.

Before Enders, Weller, and Robbins successfully propagated poliovirus in non-nervous tissue, scientists had assumed that polio was a disease of the nervous system. According to J. R.

Paul, "For a generation, starting with Flexner's early experiments in monkeys, scientists had been wedded to the idea that strict neurotropism was an outstanding property of polio." Landsteiner and Popper had shown in 1908 that polio was an infectious disease when they were able to create polio in monkeys by injecting them with virus taken from the spinal cord of a child who had died from the disease. The virus was then passaged from monkey to monkey in nervous tissue, and eventually to rats and mice. In 1936, Albert Sabin and Peter Olitsky became the first to culture poliovirus in vitro, growing the virus in test tubes of tissue from the human nervous system. Despite researchers' success in inducing polio in animal models, progress in developing a vaccine was stymied by the fact that no one could imagine injecting tissue from the nervous system into human beings. What if vaccine recipients developed encephalitis, a brain inflammation, or some other nervous system disease?

Enders, Weller, and Robbins paved the way for an understanding of the true nature of polio—that it was an intestinal disease that sometimes (not always) progressed to the nervous system. The understanding of polio as an intestinal disease also explained the sudden appearance of epidemic polio in 1916. Polio was found to be a disease of affluence, with epidemics arising from the increasingly hygienic practices of American society. Before the early decades of the twentieth century, human beings tended to live in relative squalor, living and breathing in a stew of microorganisms, one of which was poliovirus. Most people were probably exposed to poliovirus in early infancy via feces and hand-to-mouth contact. "Put simply, paralytic polio was an inadvertent by-product of modern sanitary conditions," Jane S. Smith writes in *Patenting the Sun: Polio and the Salk Vaccine*. "When people were no longer in contact with the open sewers and privies that had once exposed them to the polio virus in very early infancy, when paralysis very rarely occurs, the disease changed from an endemic condition so mild that no one even knew it existed to a seemingly new epidemic threat of mysterious origins and terrifying, unknown scope." By the early decades of the twentieth century, children were no longer commonly exposed to the virus, which had until then produced mild asymptomatic infection followed by natural immunity to the disease. When an adult or child who had never been exposed to the virus and thus never developed natural immunity encountered the three virulent strains identified by Salk and other polio researchers in the forties, they became very sick, and the disease often progressed to its final stages—paralysis and death.

By 1952, Jonas Salk and his team were producing large amounts of poliovirus by growing it in tissue culture. "Before, polio researchers had to infect individual monkeys, coddle the animals until they achieved just the right degree of sickness, and then kill them to grind up their spinal columns and take the live virus within," says Smith. "A single monkey could provide enough virus to make only a few doses of vaccine, and there weren't enough monkeys on earth to produce sufficient virus for large-scale commercial production of vaccine. Now, however, laboratory workers would be able to prepare racks of tissue culture tubes, each filled with a murky broth of tissue finely minced and suspended in a nutrient medium, 'seed' the virus in the tubes, incubate them at a temperature between 96.8 and 98.6 degrees Fahrenheit, and 'harvest' the entire crop at once."

Monkeys were still used, but fewer were needed, the process was much faster, and far more virus could be cultured from a single animal. This development was to signal an increasing reliance on in vitro studies in biomedical research. Henceforth, tissue culture in its many forms (including organ and cell culture) was to assume a place beside animal research and clinical (human) studies as a key element in the biologist's toolbox. The search for a polio vaccine is one of the first significant examples of the manner in which these three methods work together in modern research to help scientists solve the puzzle of disease and cure. Experiments on monkeys, rats, and mice were followed by tissue culture studies, which led to the development of a vaccine, which was then tested in monkeys before Salk injected himself, his coworkers, and his family.

"I look upon it as ritual and symbolic. You wouldn't do unto others that which you wouldn't do unto yourself," Salk commented afterward about this early self-testing. He then requested permission to try the vaccine on the inhabitants of the Polk School and the Watson School, preparing consent forms that were reviewed by the lawyers at the National Foundation for Infantile Paralysis. The results of the Watson and Polk trials were published in the *Journal of the American Medical Association* in March 1953. Buoyed by a favorable editorial in the same journal, Salk continued to vaccinate children in the Pittsburgh area. By the end of the year, approximately five thousand people had been vaccinated.

In 1954, the foundation's Vaccine Advisory Committee finally approved the large national field trial in which hundreds of thousands of U.S. schoolchildren participated in a double-blind study. Blood was collected from the children both before and after injections (the Salk vaccine consists of a series of three injections) to determine antibody levels, which indicated developing resistance to the virus. That year 441,131 children received the vaccine, with another two groups serving as controls: 201,229, who received placebo injections, and 1,063,951, who were given nothing. All were carefully monitored to see who developed polio that summer.

ANN ARBOR, 1955

JOHN ENDERS, who had won the Nobel Prize in 1954 for proving that poliovirus could be cultivated in non-nervous tissue, chose not to attend the one-day conference at the University of Michigan in Ann Arbor, where the results of the field trial of the Salk vaccine were to be announced. He and Albert Sabin, another prominent polio researcher, had always been opposed to Salk's approach. They favored a weakened live virus vaccine rather than the killed (inactivated) virus method used by Salk. Enders also predicted to colleagues that the scientific meeting would be a media circus.

April 12, the anniversary of the death of Franklin Roosevelt, was chosen as the day the results of the field trial would be revealed. The choice of date itself indicated that this was to be no ordinary scientific meeting; all the emotion and fanfare that the national foundation had used to fund polio research over the past two decades would fuel this meeting as well. National and international media descended en masse on Rackham Hall, site of the conference. When the university's head of public relations stepped off the elevator at 9 a.m. with the press release announcing the results of the vaccine trial, he was mobbed by more than 150 reporters who grabbed the release from his hands and began shouting "It works!" before dashing to the phones to dictate their stories.

The scientific report announcing the vaccine's safety and efficacy was read later that morning by Thomas Frances, Salk's former mentor, an esteemed epidemiologist and virologist at the University of Michigan who had run the field trial and crunched the numbers that revealed the vaccine to be 60–70 percent effective against Type I poliovirus and over 90 percent effective against Types II and III. Frances also noted that the vaccine was 94 percent effective against bulbarspinal paralysis, the most severe form of polio, which resulted in respiratoratory paralysis and death. A few hours after the meeting concluded, a fifteen-person committee of virologists chaired by William Workman of the Laboratory of Biologics Control of the National Institutes of Health unanimously recommended licensing the vaccine. The committee's recommendation was immediately accepted by Oveta Culp Hobby, the secretary of health, education, and welfare, who signed the papers licensing the vaccine that afternoon.

The outpouring of emotion that greeted this announcement was unprecedented in biomedical history. Parents throughout the United States raced to have their children vaccinated before the start of summer, polio season. Newspapers around the world carried headlines announcing that polio was vanquished. Store-fronts were decorated with signs saying THANK YOU, DR. SALK, and the scientist himself was inundated by letters of thanks and con-tributions for further research. On April 22, Salk received a spe-cial citation from President Eisenhower at the White House. Less than one week later, the party was interrupted when a vac-cinated one-year-old child in Chicago developed paralytic polio. This first case of vaccine-induced polio was followed by five oth-ers in California, with each child developing paralysis in the arm that had been vaccinated.

Panic ensued. It soon became clear that the vaccine adminis-tered to each paralyzed child had been manufactured by the Cut-ter Company of Berkeley, California, one of six pharmaceutical companies licensed to manufacture Salk's vaccine, although only two had produced the vaccine used in the field trial. Investiga-tions by the Communicable Disease Center, an arm of the U.S. Public Health Service based in Atlanta, charted the course of the outbreaks and soon revealed that there were bits of live poliovirus in the Cutter vaccine, virus that should have been inactivated during the manufacturing process.

By April 27, the virologists who had licensed the vaccine were conferring on possible responses. Should they continue the pro-gram as it was and hope for the best? Should they suspend the entire vaccination program? Should they continue the program but take the Cutter-produced vaccine off the market? The scien-tists could not agree on the best course to follow. The decision was left to the surgeon general of the United States, Leonard Scheele, who telegraphed the president of the Cutter Company, proposing that the company voluntarily cease distribution of its vaccine. The company did so. Nonetheless, 400,000 children had received the Cutter vaccine between April 12 and April 27. Of that number, 79 children developed polio, with 125 other individuals—family members and playmates—infected through them. Three-fourths of those infected were paralyzed, and eleven died. "When compared to the 5,000,000 people who had received the vaccine, the 250 Cutter cases [there were more after April 27] and another 115 cases among vaccinated children seemed like a very good record for the vaccine," wrote Aaron E. Klein in *Trial by Fury: The Polio Vaccine Controversy*, adding that "these beautiful statistics were no comfort to the parents of the stricken children, or to Salk, the Foundation, or the Cutter company. Lawsuits were filed and took years to be settled. The early surge of demand for the vaccine was followed by a period in which parents had to be urged to vaccinate their children and themselves. Most eventually did so, and by 1960 only 2,525 cases of paralytic polio were reported in the United States."

In 1961, Albert Sabin produced a single-dose oral vaccine, which was field-tested in the Soviet Union since most Americans had by then received the Salk vaccine. Within months of the vaccine's licensing in the United States, sixty people who had received it developed vaccine-associated polio. Nonetheless, by the mid-1960s most people in the United States were receiving the one-dose oral Sabin vaccine and not the three-dose injectable Salk vaccine. In 1965, only sixty-one cases of paralytic polio were reported in the United States. By 1990, an average of eight cases a year were being reported, most of them associated with the Sabin vaccine. Polio had been conquered.

"Paralytic poliomyelitis is a phenomenon of the twentieth century, a disease that went from epidemic appearance to near-extinction in fifty years," writes Jane S. Smith in *Patenting the Sun*. The numbers she cites are shocking: "Twenty-seven thousand people were paralyzed in the epidemic of 1916, and six thousand died. There were twenty-five thousand cases of poliomyelitis reported in 1946, fifty-eight thousand in 1952, thirty-five thousand in 1953. In 1957, the first year the Salk vaccine was used, five thousand cases were reported. By 1960, a year before the Sabin oral vaccine was introduced, the number of cases had already dropped to three thousand."

Barely a generation after the development of the polio vaccine, memories of the terror inspired by this disease, which had disabled or killed hundreds of thousands, had grown dim. But the heroic vision of a biomedical science able to vanquish such a formidable foe survived. Given enough money (and enough animals), scientists could overcome any disease, it seemed. But the paradox of polio, and the threat that paradox implied, lingered long after the cheering over the development of a vaccine had subsided. Unlike typhus, plague, and many of the other infectious diseases that had long shadowed mankind, epidemic polio was not bred in filth, and better hygiene could not eradicate it. The virus lived in close proximity to human beings and needed only a slight alteration of circumstances—indoor plumbing and modern sanitary practices—to metamorphose from a benign neighbor to a killer. But this salient fact was largely overlooked in the general celebration. In the immediate aftermath of the development of the polio vaccine, scientists like Salk had the pop-culture appeal of Superman, mild-mannered but all-powerful and able to overcome any adversary.

As the fifties waned, however, the man of steel began to seem increasing irrelevant to some, his imperturbable authority and invulnerability somehow threatening, a kind of friendly fascism. A new sort of culture hero was taking shape, a more complicated figure, powerful but not invulnerable, a creature of darkness and the night who rose from the shadows to defend the powerless in a nightmarish city awash in corruption and greed. As decades change, so do heroes. This comic book defender of justice was nearly as old as Superman, but he was a champion of a different kind, and it took decades for him to find as wide an audience.

Identified with an animal, he fought a long-running battle with villains who, rather than being redeemed by science, had been corrupted and deformed by it. His adversaries were in some cases scientists or former scientists, who like Robert Louis Stevenson's character, Henry Jekyll, had been damaged either physically or psychologically by their own experiments. The villains had names like Mr. Freeze, Scarecrow, Two-Face, and Poison Ivy. This hero was a dark knight, and though he didn't reach his peak of popularity until the eighties and early nineties, by the mid-sixties he had already surpassed Superman in sales, becoming the pop-culture harbinger of a new style of crusader—by day a respected member of the elite, and by night an avenging angel, swooping down on his enemies like a bat, cloaked in darkness.

—S I X—

THE NEW CRUSADERS

Animal liberation is also human liberation. Animal libera-
tionists care about the quality of life for all. We recognize our
kinship with all feeling beings. We identify with the powerless
and the vulnerable—the victims, all those dominated, oppressed
and exploited. And it is the nonhuman animals whose suffering
is the most intense, widespread, expanding, systematic, and
socially sanctioned of all. What can be done? What are the pat-
terns underlying effective social struggles?

—Henry Spira,
"Fighting for Animal Rights:
Issues and Strategies," 1983

NEW YORK CITY, 1974

H ENRY SPIRA was all fired up. The forty-seven-year-old English teacher had just taken an adult extension course at
New York University offered by an Australian philosopher named
Peter Singer. Singer had recently published a review of *Animals,*
Men and Morals, a collection of essays that challenged prevailing
notions of the human-animal relationship, in the *New York*
Review of Books. Spira hadn't seen the piece when it was published, but he came across an attack on the philosopher's views in
the leftist *National Guardian* and found himself agreeing with
Singer, not the author of the attack. When an opportunity arose
later that year to take a course with the visiting professor from
Australia, Spira jumped—and landed in exactly the right place at
exactly the right time.

By the time he met Peter Singer in 1973, Henry Spira had a long history of involvement in various campaigns for workers' rights, human rights, and civil liberties. Born in Belgium in 1927, Spira had a restless childhood. Shuttled from relative to relative throughout the late 1930s because of the family's emotional and financial difficulties, he learned to adapt quickly to different environments. Early on he developed a preference for the under-dog and a sensitivity to suffering. He and his family were Jews, and although they left Europe before the Holocaust, moving to Panama in 1938 and New York City in 1940, he was with his mother and sister in Hamburg on *Kristallnacht,* when Germans smashed the windows of Jewish shops and burned synogogues. Safe in New York City during the war, he overheard the conver-sations of the adults, the anguished discussion of horrors beyond comprehension. He felt fury—why were the adults just talking, why didn't they do something instead of moaning and kvetching? He was a child and he was powerless. But he promised himself that when he was an adult he would not be one of those who complained about injustice yet did nothing. He also became an atheist, concluding that continued belief in a just and merciful God after the Holocaust "defied common sense."

At sixteen, Spira left home, and a year later he joined the mer-chant marine. He traveled the world, immersing himself in the revolutionary writing of Marx, Trotsky, Plekhanov, and Emma Goldman. Screened off the ships as a security risk in 1952, Spira was drafted into the U.S. Army that same year and sent to Berlin as a troop information educator. He was discharged in 1954 and went back to work in the merchant marine. He soon learned about rampant corruption in the National Maritime Union (NMU), in which a boss named Joseph Curran was bilking sea-men out of benefits and threatening anyone who dared challenge his one-man rule of the union. Spira began editing *The Call for Union Democracy*, an insurgent paper that challenged Curran's regime, writing articles to expose the luxurious lifestyle enjoyed by the union president while he cut seamen's pensions and increased their dues.

"What can be done within a job trust such as the National Maritime Union, where Joseph Curran (its first and only president) and his apple polishers feast and fatten themselves by milking the working seamen?" Spira wrote in 1969. "Where seamen are bound, gagged and helpless, when all their rights have been taken away, when a labor organization has been transformed into a racket for the enrichment of one man and his 'yes men' and then those who would return the union to the membership are forced to utilize the legal process to assert their constitutional rights? Publicity is necessary for physical protection. When hoodlums know that the press is watching, the terrorizing of opportunists tends to be restrained."

The NMU insurgents were partly successful when Joseph Curran retired, albeit with a $1 million bonus. But by then Spira had expanded his focus to other rights struggles, traveling in the South and writing about civil rights activism, FBI surveillance of U.S. citizens, and workers' rights. He kept up his contacts in the NMU and among fellow leftists in New York City while he went back to school, got a degree, and started working as an English teacher in the public schools.

By the time Spira took Singer's class, in 1974, he had begun to think about animals. A friend had given him a cat, his first pet, and after reading Singer's essay he began to think about the paradox of the human-animal relationship as he found himself cuddling and petting one mammal and sitting down to dine on another. But it wasn't until he took Singer's class at NYU and heard the philosopher expound concepts from his manuscript in progress, *Animal Liberation*, that Spira made the connection between his intuitive discomfort and the systematic oppression he had battled throughout his life.

In an interview published in the magazine *Animal Liberation* in 1989, Spira told Singer: "I think that what your essay did for me, what the class did for me, was to put the whole issue of animal rights first within the context of liberation movements and secondly, put it on a rational basis that can be defended in public debate on its own merits without reference to whether one does or doesn't like animals. What came out of the course was that animals are being harmed on a massive scale, and that it's wrong, it's an injustice. Clearly, the next thing is, if it's wrong, what are we going to do about it?"

What are we going to do about it? It was the question that had haunted Spira down through the years whenever he was confronted with injustice. But by 1974, after years of work and struggle for change, he knew exactly what to do. On the last day of class he invited anyone interested in following up on the issues discussed in class to meet at his apartment the following week. There they could talk about next steps, about translating outrage into action. The person Spira most wanted to attend was a tall, thin psychiatrist whom Spira had pegged immediately as brilliant but a bit odd, a thoroughgoing skeptic and independent thinker like himself. The doctor's name was Leonard Rack. We'll need a scientist, Spira told himself, someone on the inside, like the NMU officials who had secretly collaborated with the insurgents, just as disgusted by the union's corruption even though they were part of the system. Although Rack had attended only one of Singer's classes, Spira contacted him anyway. They talked about the class, and Spira told him that a few of the students were interested in looking at some research reports to see if Singer was right about animal abuse in laboratories. Since none of them were physicians or scientists, they would probably need help understanding the purpose of the experiments and deciphering the jargon. A longtime ethical vegetarian, Rack was intrigued and agreed to attend the meeting. Spira had his insider.

IN AN AGE of heroic science—man on the moon and antibiotics, plastics and heart transplants—the vast majority of the public respected and trusted scientists and seemed not to question their need or right to experiment on animals, even though the numbers of animals used in research throughout the world had reached an all-time high. Surveys conducted by the International Committee on Laboratory Animals and independent scholars revealed that between 1959 and 1975, laboratory animal use more than doubled in most countries, with far greater increases in some. "Japan's use of laboratory animals increased from 1.6 million in 1956 to 13.2 million in 1979. Use in the United States had increased 18 million in 1957 to at least 51 million in 1970. Estimates of total worldwide use ranged from approximately 100 million to as many as 225 million," according to Andrew N. Rowan in *Of Mice, Models and Men: A Critical Evaluation of Animal Research.*

In the United States, federal appropriations for biomedical research had similarly skyrocketed. Most of this research was funded by the National Institutes of Health (NIH), which grew out of the Hygienic Laboratory set up at the U.S. Marine Hospital on Staten Island in 1887. In 1948, the Hygienic Laboratory was transformed by an act of Congress into the National Institute of Health, and by 1975 it had become a research giant, sponsoring intramural research at eleven separate institutes on the main NIH campus in Bethesda, Maryland, and funding extramural research at a host of universities and research institutes throughout the United States and abroad. The costs of that transformation were substantial: the NIH budget rose from $29 million in 1948 to $1.4 billion twenty years later.

The money produced results. "By 1983 over 80 Nobel laureates in Chemistry and Physiology or Medicine had conducted at least part of their research with NIH support," Victoria Hardin comments in her history of federal biomedical research policy. "Inspired by the outstanding record of NIH scientists, Wilbur J. Cohen, then secretary of health, education, and welfare (HEW), declared in 1968: 'The National Institutes of Health are a brilliant jewel in the crown of HEW.'" Over the next decade, important discoveries continued to be announced by U.S. researchers, most of them funded in large or small part by the NIH and supported by the tax dollars of the American people.

In the United States, antivivisection seemed to many a dusty relic of a bygone era. In the late sixties and early seventies, some people still remembered polio and lethal influenza epidemics and were grateful for the protection conferred by science. Public condemnation of the Tuskegee Syphilis Study in 1972 led to the creation of institutional review boards (IRBs) in all research institutions to review protocols involving human subjects, and a few philosophers who had begun to call themselves bioethicists began to debate the concept of informed consent, but these developments did not have much effect on animals. Although organizations like the Animal Welfare Institute in the United States and the Universities Federation for Animal Welfare (UFAW) in Great Britain still worked quietly to improve conditions for laboratory animals, it was an uphill battle, particularly in the United States, where the scientific community reacted with knee-jerk ferocity to any attempt to oversee or regulate biomedical research. Even the creation of institutional review boards to monitor research on human subjects was met with resistance by many who argued that IRBs constituted an attack on academic freedom and scientific integrity.

In the United Kingdom, where the 1876 Cruelty to Animals Act had created a federally administered system of licensing and regulation, things were a bit different. Scientists were accustomed to government visitors in their laboratories, even if their inspections were less than thorough, and animal defenders like Brigid Brophy and Air Chief Marshal Lord Dowding and his wife, Muriel, kept the antivivisectionist ideal alive. "I firmly believe that painful experiments on animals are morally wrong, and that it is basically immoral to do evil in order that good may come— even if it were proved that mankind benefits from the suffering inflicted on animals," Dowding testified in the House of Lords on July 18, 1957. Although British antivivisectionists like Dowding and his wife (who was a Theosophist, spiritualist, and vegetarian like Anna Kingsford) believed that the 1876 law was seriously flawed, the Cruelty to Animals Act nonetheless created a mindset that accepted the need for a certain degree of government scrutiny in the conduct of research.

Moreover, in 1954, the Universities Federation for Animal Welfare had commissioned a scientific study of humane technique in experimentation under the direction of the immunologist Sir Peter Medawar. Christine Stevens contributed a small grant from the Animal Welfare Institute to partially support the work. Soon after the start of the study, she traveled to England and met the two scientists who were to conduct the research, visiting laboratories and compiling data. W. M. S. Russell, an Oxford-trained zoologist, and Rex Burch, a microbiologist, were young and enthusiastic about the project, and both the UFAW and the AWI had high hopes that the study would change the way laboratory animals were used and cared for, based on its solid scientific hypothesis—that effective research demanded high standards of animal care.

Russell and Burch's book, *The Principles of Humane Experimental Technique*, was published in 1959 by Methuen and Company of London. A few thousand copies were sold in England and even fewer in the United States, but for the most part the book was ignored. Stevens was disappointed. Her visits to U.S. laboratories still revealed dreadful conditions in some—filthy animal quarters; infestation of ticks, roaches, and other insects; wild rodents making their way into laboratories and spreading disease among the laboratory rats and mice. Improvements were desperately needed, Stevens and her supporters contended. Prodded by AWI's reports and other criticism, the National Institutes of Health in 1965 commissioned its own study, to be carried out by David N. Ruth and Robert W. Cox. In response to the researchers' request for information, AWI submitted documentation detailing its members' observations during laboratory visits—cages so small that animals could not stand or lie normally, failure to administer analgesics after surgery or to euthanize suffering animals, repeated use of the same animal for painful procedures, and more.

Ruth and Cox produced a 210-page report, *The Care and Management of Laboratory Animals Used in the Programs of the Department of Health, Education and Welfare,* which was immediately squelched by the NIH hierarchy because of its frank assessment of conditions similar to those described by AWI and inclusion of AWI's charges in its appendix. Soon after, testimony in the United States Congress by British scientists on the success of regulation in Great Britain under the Cruelty to Animals Act of 1876 was abruptly suspended. The power of the U.S. research community, which recoiled from any discussion of regulation by outsiders, seemed too strong to overcome. Stevens and her small band of reformists were outmaneuvered, despite the fact that she herself now had some powerful friends and allies. Roger Stevens, Christine's husband, had been appointed the first director of the Kennedy Center in Washington, D.C., and the couple were on good terms with many important political figures, including Hubert Humphrey and Robert Dole, both of whom supported animal protection legislation. Although Stevens had successfully lobbied for the passage of the Humane Slaughter Act, signed into law on August 20, 1958, every laboratory animal reform bill that her congressional supporters introduced in the House between 1960 and 1965 was killed by the powerful opposition of organized research interests.

Then, in July 1965, AWI got a call from a woman named Fay Brisk, who had long worked for reforms among the Pennsylvania dog dealers who supplied animals to laboratories. Brisk told Stevens that a Pennsylvania family had seen photographs in a local newspaper of their missing dog, Pepper, being loaded onto a dog dealer's truck. The family had then set off on a two-state chase in search of their missing pet and wound up on the grounds of a New York dealer named Neresian, who refused to let them see the animal. Stevens called her contacts in Congress and they attempted to intervene, but Neresian refused even the request of his district congressman to visit the property. Meanwhile, the Pennsylvania dog dealer identified in the newspaper article admitted that he had taken all the animals in the photograph (including the one the family identified as Pepper) not to Neresian but to the laboratories at Montefiore Hospital in New York City. The hospital admitted that the dog in question had died on the operating table and been incinerated.

The arrogance and lack of cooperation exhibited by the dog dealers infuriated Joseph Resnick, the congressman who had worked to help his constituents locate their missing pet. On July 9, 1965, he introduced a bill in the House of Representatives requiring that dealers in dogs and cats, as well as the laboratories who purchased the animals, be licensed and regularly inspected to ensure humane treatment. Senators Magnuson, Clark, and Brewster cosponsored a similar bill in the Senate. The National Society for Medical Research and the federal research agencies attacked the bills immediately, insisting that they were attempts by antivivisectionists to cut off the supply of animals to medical laboratories—despite the fact that the antivivisection organizations did not support the bills, as they aimed only to regulate, not abolish, animal experimentation. The Resnick and Magnuson bills would likely have died in committee if it hadn't been for Henry Luce and *Life* magazine.

In the midst of the hearings on the Resnick bill, which began in September 1965, Stevens sent Luce a stack of photographs taken at dog dealers' properties and laboratories by AWI staff. Intrigued, Luce directed a reporter and photographer to look into the story. On February 4, 1966, a front-page story entitled "Concentration Camp for Dogs" was published in *Life* and immediately set off a storm of public outrage. The pathetic photographs of emaciated animals in filthy, overcrowded pens infuriated animal lovers across the country, and they demanded passage of the Resnick bill. Mail poured into the offices of the magazine as well as congressional offices. Opponents of the Resnick bill were backed into a corner and fought like most creatures placed in a similar situation. The National Institutes of Health and the Department of Health, Education and Welfare requested first that laboratories be exempt from the legislation and then proposed a counter-bill that mandated regular inspections by a one-year-old organization called the American Association for the Accreditation of Laboratory Animal Care. AAALAC was staffed and run by animal researchers and had the NSMR's stamp of approval. Despite these machinations, the Laboratory Animal Welfare Act was passed by Congress on August 24, 1966, with enforcement powers placed in the hands of the Department of

Agriculture. Although the Act did not empower the secretary of agriculture to oversee animal research itself, and authorized only the regulation of "transport, sale, and handling of animals" prior to their arrival in laboratories, it was viewed by its supporters as a victory. Slowly, a legislative apparatus protecting laboratory animals was being put in place.

NEW YORK CITY, 1976

"IF I WERE in their shoes, what would cause me to change my behavior?"

That was the question Henry Spira asked himself when he thought about the American Museum of Natural History. He and the other activists in his group had discovered through Freedom of Information Act searches that scientists working at the august institution a few blocks from his apartment on the Upper West Side of New York City had been conducting experiments on the sexual behavior of cats for nearly two decades. Lester R. Aronson, curator of the museum's department of animal behavior, and his assistant, Madeleine Cooper, had been funded by the National Institute of Child Health and Human Development to study behavioral changes in cats deprived of different kinds of brain function. Despite long study and over $400,000 in federal grants, the work had resulted in few publications and very few citations in the scientific literature.

Spira's science adviser, Leonard Rack, had told Spira and the rest of the group that in his opinion the cat sex studies had little scientific value, and Spira intuitively felt that the museum was a perfect target to initiate his campaign to publicize the cruelty and waste of money and animal lives in research. So, beginning in the summer of 1975, he and his friends from Singer's class began spending weekends outside the museum, protesting and passing out flyers. Later that year he began to write a series of articles on the experiments for *Our Town*, a Manhattan weekly whose publisher-editor was a cat lover. On May 23, 1976, the paper published a long piece charging that the experiments were both cruel and scientifically useless. In the article, Spira quoted from the researchers' grant applications and research reports to bolster his argument:

"On the Fifth Floor of the Museum's Education Building, cats and kittens are deliberately blinded—both eyes cut out. Their hearing and sense of smell destroyed by slicing into their brains. Their sense of touch deadened by cutting nerves in their sex organs. The Museum's experimenters score the mutilated animals on their sexual performance. Some male cats are subjected to a terminal electro-physiological experiment, forced into a rack where the penis is stimulated with filaments. The cats scream and the Museum puts up a 'sound retarded' room (M5285, p. 9). Cats are driven out of their minds by pain and stress and the Museum staff confines them into 'special testing pens and transfer cages as a safety device. We expect that certain animals . . . will be difficult or impossible to handle in an ordinary manner' (HD00348-14, p. 4). And the taxpayers have been paying for this 'research' during the past 17 years. In the last reported year, 74 cats and kittens were cut up alive. To date, $437,000 has been appropriated through Grant HD000348 from the National Institute of Child Health and Human Development, Bethesda, Maryland 20014."

"YOUR MONEY IS PAYING FOR TORTURE AT THE AMERICAN MUSEUM OF NATURAL HISTORY," blasted the flyers that Spira and his growing cadre of protesters passed out on the sidewalk in front of the museum during the summer of 1976. "The cat experiments, once secret, have gained national media attention, including NBC-TV features, network news programs, major articles in the *Christian Science Monitor, New Orleans Times-Picayne, Philadelphia Inquirer, Miami Herald, Chicago Sun Times, New York Times, Daily News, New York Post*, the *Congressional Record*, and upcoming articles in *Science* and campus papers . . . STOP THE TORTURE, CLOSE THE LABS, SAVE THE CATS, BOYCOTT THE AMERICAN MUSEUM OF NATURAL HISTORY." The protest captured the attention of legislators such as New York congressman (later mayor) Ed Koch, who visited the laboratories and then wrote a letter to the secretary of health, education and welfare, commenting, "While I am not prepared at this moment to label this kind of experimentation as Nazi-like, it does recall the barbarities of the Nazis."

For eighteen months, the protesters sat in front of the museum every weekend passing out flyers and carrying signs. "We used a variety of tactics to maintain the energy and pressure," Spira noted in *Strategies for Activists*, a primer on activism he self-published in 1997. "For example, admission to the Museum is by voluntary donation, but with a suggested fee. We gave visitors a penny as they approached the entrance, asked them to ignore the Museum's suggested fee, and upon admission give only the penny. Then, the Museum threatened to arrest anyone entering the Museum with a protest sign. To counter this, Ed Kayatt (publisher-editor of *Our Town*) supplied buttons and produced T-shirts with slogans similar to picket signs. These were worn by hundreds of protestors throughout the Museum. The Museum could hardly ask us to take off our T-shirts."

In October 1976, the campaign received an unexpected boost when a story about the museum's troubles appeared in *Science*, the much-respected publication of the American Association for the Advancement of Science. The article, authored by *Science* staffer Nicholas Wade, provided an unexpectedly balanced presentation of the controversy and a prophetic analysis of the research community's vulnerability. "The museum's plight carries a warning for other institutions whose experiments with animals are susceptible to being made the focus of public passions," Wade wrote. "The animal rights groups are particularly well-informed about the cat study because, through the Freedom of Information Act, they obtained all the investigator's grant applications from the National Institutes of Health. Second, the issue of animal rights has been taken up recently by several young philosophers whose writings have injected a new intellectual vigor into the movement. The animal rights groups believe that there is a historical trend in their favor which goes from minorities' rights to women's rights to animal rights. The attack on the Museum of Natural History is just the first shot in what they hope will be a broader campaign."

Under the provocative subhead "Sadism Frowned Upon," Wade wrote that "while most researchers doubtless respect the interests of their animals as much as possible, the codes of practice governing animal experimentation do not concede that animals have any rights whatever that should weigh against the purposes of the experimenter. . . . Just as the researcher is not required or formally encouraged to make animals count for anything in the design of his experiment, so the peer review system makes no formal attempt to balance the worth of an experiment against the interests of the animals whose lives it would take."

Wade ended the article with a quote from Peter Singer's *Animal Liberation*: "Surely one day, our children's children, reading about what was done in laboratories in the 20th century, will feel the same sense of horror and incredulity at what otherwise civil people can do that we now feel when we read about the Roman gladitorial arenas or the 18th century slave trade." He then added a coda of his own: "The projection may sound far-fetched, yet history teaches that only fashions in clothes change faster than fashions in ethics."

In December 1977, eighteen months after the protests began, the Museum of Natural History halted the experiments. The principal investigator retired, and the museum's laboratories were closed. In a follow-up article in *Our Town*, describing the outcome and implications of the campaign, Spira quoted a memo written by Stephen Toulmin, a member of the National Commission for the Protection of Human Subjects, which stated that "the same climate that has been unfavorable to biomedical research involving human subjects is also unfavorable toward biomedical research involving animals. The recent shindig at the Museum of Natural History in New York may be, in this respect, an indication of a difficult phase that animal research workers are going to have to live through in the years ahead."

Spira heartily agreed. "We may well be entering the decade of animal rights," he wrote. "We've won the first battle. But as you read this, there are millions of animals in labs. And lab animals never have a good day. So, the fight goes on. And nothing, not even a multi-billion dollar animal research cartel with powerful political connections, can stop an idea whose time has come— that we must treat other feeling creatures the way we'd want to be treated were we in their place."

MARYLAND, 1981

ALEX PACHECO took out his camera and began taking pictures. He photographed the rusty cages, which were crusted with feces; he photographed the dirty, disorganized laboratory and the signs of insect and rodent infestation. But mostly he took pictures of the monkeys and the bloody stumps on their hands. They looked a lot like human hands, with some of the fingers gnawed to the bone. Later, Pacheco, who had posed as a student volunteer to gain access to the facility, brought in five veterinarians and primatologists—sympathetic scientists who studied monkeys, chimpanzees, and apes—sneaking them in at night after all the laboratory workers had gone home. He wanted them to see the conditions in the laboratory and to confirm his own observations about the filth and the lack of basic care for the animals he had seen there, according to an account he cowrote with Anna Francione and published in the book *In Defense of Animals*, a collection of articles edited by philosopher Peter Singer. The local Humane Society had received complaints about this lab before and had investigated the principal investigator, Edward Taub, but nothing had come of it. Pacheco was gratified when all five of the medical people signed affidavits testifying to the poor conditions of the animals and the facilities in the lab.

Pacheco turned the photographs and affidavits over to the Montgomery County Police, and on September 11, 1981, a group of officers showed up at the Institute for Behavioral Research with a search warrant. They removed seventeen monkeys from the lab and charged its director, Edward Taub, with cruelty to animals. Pacheco and his associates at the newly formed People for the Ethical Treatment of Animals (PETA) alerted every reporter and news organization they could access. The story was soon carried on wire services and spread throughout the country and overseas. Americans were appalled, and the National Institutes of Health undertook its own investigation to determine whether the laboratory was complying with Public Health Service policy on animals. At that time, Taub had been funded by the National Institutes of Health for nine years. In October, the NIH found the laboratory and its director in violation of Public Health Service policy on a number of counts and withdrew funding.

"Such a decision by NIH was not made lightly. Indeed, it was well-recognized by Thomas E. Malone, Acting Director, and by other senior NIH officials that to stop funding a project because of animal welfare concerns would cause bad feeling, even ire, within the biomedical community," comments former NIH researcher and animal welfare author Barbara Orlans in her 1993 book *In the Name of Science: Issues in Responsible Animal Research*. "However, they believed it would have been impossible for NIH to maintain credibility with the public and Congress without this action." Taub had been using a procedure known as deafferentation, opening the spinal cords of monkeys and slicing the nerves leading to their arms and legs. Scientists had long believed that limb function was permanently lost when nerves were damaged by stroke or by accidents. Taub was trying to prove otherwise—that it was possible for nerves to be regenerated. The effects of deafferentation on animals were well-known to researchers. As Barbara Orlans has noted, "Deafferented monkeys are notoriously difficult to take care of because they treat the deafferented limb as a foreign object, and frequently chew off fingers. . . . Among the group of twelve deafferented monkeys (in Taub's laboratory), thirty-nine digits were missing or deformed. Photographs of the animals' deformed hands raised strong public emotions."

At Taub's trial in November, veterinarians and fellow physiologists testified in Taub's favor, verifying the scientific merit of his work and denying that conditions in the lab were substandard or that Taub had failed to provide proper care for the animals. Nonetheless, Taub was found guilty of providing inadequate care to six of the animals. (The conviction was later reduced to one count of cruelty, and then was overturned in 1983 by the Maryland Court of Appeals, which ruled that the state's anticruelty statute did not apply to federally supported scientific research.) Edward Taub's conviction sent shock waves through the research community. If a respected physiologist, whose work was scientifically valid according to all the standard criteria, could be arrested on cruelty charges, what chilling effect would that have on research in the United States? Coming on the heels of Henry Spira's campaign against the Museum of Natural History and his subsequent successful effort to repeal the Metcalf-Hatch Act in New York State in May 1979, the Silver Spring monkey case became a cause célèbre among both animal rights advocates, who could feel the momentum of their movement growing day by day, and the research establishment, which for the first time in eighty years was facing a challenge that could not be overcome by a few days of expert testimony.

Animal rights lore indicates that the Silver Spring monkey case had another effect. According to a book written by PETA cofounder Ingrid Newkirk, *Free the Animals: The Inside Story of the Animal Liberation Front,* one of the officers who heard the call for the raid on the Institute for Behavioral Research on her squad car radio became so distraught over the monkeys condition that she helped a group of activists kidnap them after a Maryland judge agreed with Taub's claim that the monkeys were legally his property and should be returned to him. One month later, Newkirk wrote, this woman flew to England to meet members of an underground group called the Animal Liberation Front (ALF), who were becoming notorious for raiding research laboratories, stealing animals, vandalizing and burning offices, and destroying documents and files. The U.K. group's founder, Ronnie Lee, had served time in prison in 1974 after he was caught during a lab break-in, and had been imprisoned again in 1977 following another break-in. Working through a British animal rights group, this former police officer, whom Newkirk refers to as "Valerie" in *Free the Animals*, had contacted the ALF and was then trained by the group. Returning to Maryland in the fall of 1982, she began recruiting soldiers for her war and founded the American wing of the Animal Liberation Front, an organization that was soon to begin disturbing the sleep of researchers throughout the country.

"VALERIE IS TOTALLY FICTITIOUS," attorney Jo Shoesmith, a longtime activist and one of the core members of the group that later became People for the Ethical Treatment of Animals (PETA), told me in a July 1998 interview. She first met Ingrid Newkirk, one of the founders of the group, in 1978, when she volunteered at the Washington Humane Society (WHS) during a summer break from college. Newkirk, at that time director of cruelty investigations for the WHS, "was very much on the street, picking up animals, responding to calls in godforsaken parts of the city," and Shoesmith accompanied her on some of these runs. The two women became friends. "In a very informal, subtle way, that was a real turning point in my life," Shoesmith said. "In my experience with her, Ingrid was never one to sit down and give you a talking-to about animal rights. A friendship developed that evolved very naturally." At the end of the summer, Shoesmith wanted to stay, but Newkirk encouraged her to return to college, telling her that an education would help her work in animal rights. Shoesmith went back to school and the two women lost contact.

After graduation two years later, Shoesmith returned to Washington and began working with the Monitor Consortium, a marine mammal protection group. "I ran into Ingrid and she said, 'We're just forming a group at George Washington University. Would you like to come?' That's where I met Alex [Pacheco]," said Shoesmith. "He was a student at GW and had the connection to the university, which gave us a place to meet." It was a small group at first, with participants ranging in age from early twenties to early fifties. Most of the people in the group had read Peter Singer's *Animal Liberation.* "That book was instrumental in putting language to a gut feeling," Shoesmith said. "It altered my life course permanently. I just knew that it was something that I was born to do."

The fledging animal rights group met monthly at George Washington University. Henry Spira took the train down from New York for a couple of meetings. "They were like therapy sessions," Spira told me in 1996 and again in 1998, "people talking about their feelings." Shoesmith agreed that this was a fair assessment. Spira said that he urged the group to translate their feelings into action, then went back to New York and never returned. According to Shoesmith, the group's first "active engagement" after the talking stage was the picketing of inhumane conditions at a poultry slaughtering plant in Washington, D.C. Soon after that, "Ingrid and Alex decided to rename the group. Ingrid told us the name at a meeting one night, but none of us really understood what she was saying because she has this wonderful accent. We thought she was saying 'Peter.' It wasn't a group decision, but of course we came in under this new name that rather mystified us."

The decision to rename the group, and the top-down conveying of the information to its members, was fairly typical of the organization's functioning, Shoesmith said. Even at that early stage, PETA was far from a democratic system: Pacheco and Newkirk were in charge, and the rest of the group fell into line. "You operate on a need-to-know basis," Shoesmith said she and the other members of the group were frequently told. "Consequently, everybody felt both included and excluded at certain times. You never knew how long things had been in the works before you found out about them." For that reason, Shoesmith isn't sure when she found out about Pacheco's work in Taub's laboratory. But she does remember his horror and disgust over the conditions in the lab. "I remember him saying how long it would take him to shower at night to get the stench off," she said, "and how completely hideous the lab was. Also how he was driven to be there despite how horrible it was to be inside the lab."

Shoesmith called what happened next "the high alert phase," when Pacheco was gathering the evidence he needed to indict Taub. In the weeks before the raid, "we were photocopying documents late at night. Ingrid and Alex brought in a photographer from Vancouver, Peter Hamilton, to take the photographs. They were so perfectly hideous. We developed a kind of perverse ability to see the public relations value of these things. At that time, the offices of PETA were in Ingrid's living room in Takoma Park. There was no living space. Things were cranked out, with no letup." The activists blew up the photographs into glossy posters that the group "plastered all over Georgetown," Shoesmith said. For months, the group worked to raise public consciousness and get the maximum publicity. "For me, the trial itself was anticlimactic," said Shoesmith. Recalling the months that Pacheco worked at the Institute for Behavioral Research and the push to publicize the conditions there, she said "there is some adrenalin, but you're always working on the edge of burning people out under those conditions."

In later years, after she and the group had parted ways, Shoesmith heard stories about the same kind of authoritarian management described by other current and former PETA employees. "By the mid eighties, there was unrest among people working there. The tone is set from the top, and we were given to understand that this was a mission. No one should be there for accolades. If you expected a pat on the back, you just weren't going to get it. There are people today who would speak with great hostility about that. . . . You're working with the wrong kind of people to be lording it over. Animal rights activists by nature are individualists, so you had the perfect recipe for unrest going on all the time."

Nonetheless, Shoesmith spoke of Newkirk (and to a lesser extent Pacheco) with great respect. "I don't think that anyone has matched Ingrid Newkirk to this day," she said. "She was well-spoken, astute, sharp as a tack. She could be very caustic. So witty, so uncompromising. The personification of the movement. Also totally accessible. A reporter could call her at five in the morning for a quote, and she'd be just as sharp as ever. We were in awe of her."

Newkirk could also be "very maternal," according to Shoesmith. "She was cooking for us constantly and was very generous in taking care of the flock, but she would confront you if you said something that you shouldn't. She was the glue that kept the whole organization together. It's hard not to admire those qualities." Shoesmith said that Newkirk and Pacheco "were an incredible team. Alex was the front guy, the one who would go and do the dirty work, put it on the line with his physical presence. They filled every niche."

In 1981, this group and other grassroots animal activists working throughout the country decided to hold what they believed to be the first modern International Animal Rights Conference. A series of planning meetings were held in Washington and New York to prepare for the conference. "In this very anarchistic group, the need developed for a general coordinator-peacemaker," said the soft-spoken Shoesmith. "And I was asked to take on the job." For the first time, the new breed of grassroots volunteer activists were going to interact with the organizational folks, people from the American Antivivisection Society and the National Antivivisection Society, the big-money groups funding the conference, to work together to define the movement's goals and develop a strategic plan.

"We met at the Francis Scott Key Hotel in Ocean City, Maryland, for three days," Shoesmith said. "The intensity level was so high that there were people who didn't sleep the whole time. I didn't sleep. By the third day, there started to be a whole lot of dissension about where to go from here, setting out a ten-year plan, what to achieve, who was going to do what, where to meld efforts and coordinate efforts nationwide, and things just broke down. People had not enough sleep and too much coffee. It was quite dismal." Shoesmith laughed. "I made a pleading appeal to the core group, standing up in front of this long table where the group was assembled. I made a plea for unity, for putting our differences aside. I sat down with tears streaming down my face. It was moving, but in the end people turned back to their own business. For about 30 seconds people put down their egos, but then . . ." She shrugged. "That kind of coordination is just not in the nature of the people nor in the nature of the cause. No one wanted to commit to a joint program when each group wanted to carve out its own niche and be guided by its own internal agenda."

THE GROWTH of the animal rights movement during the 1980s remains a relatively understudied social phenomenon. Membership in some of the older animal welfare societies had been growing slowly throughout the middle years of the century, but the great leaps in membership enjoyed by the new rights-oriented groups during the 1980s were truly astounding. The relatively moderate Humane Society of the United States, founded in 1954, added around 15,000 new members each year from 1978 to 1984. From 1984 to 1988, "it added 100,000 new members each year, to reach a total size of over half a million by 1988," sociologists James M. Jasper and Dorothy M. Nelkin pointed out in their study of the movement, *The Animal Rights Crusade: Growth of a Moral Protest*. The same trend has been noted by other scholars.

"In the first few years of the 1980's important national organizations originated, including People for the Ethical Treatment of Animals (PETA), Transpecies Unlimited, Farm Animal Reform Movement (FARM), Feminists for Animal Rights, Mobilization for Animals, and In Defense of Animals," wrote Susan Finsen and Lawrence Finsen, professors of philosophy and authors of *The Animal Rights Movement in America*. Science reporter Deborah Blum, whose 1994 book *The Monkey Wars* provides an extensive and balanced discussion of the use of primates in biomedical research, confirmed that "there are now more than 400 animal advocacy groups in the United States, claiming a combined membership of 10 million and total income approaching $50 million. The money itself tells the support for animal advocacy; it comes, almost entirely, directly out of the pockets of Americans angry about the treatment of animals."

Blum noted that the new advocates for animals were not only numerically but also philosophically estranged from their predecessors. "Out of the 1980's came a new breed of animal activists," she said, "fashioned in the image of Silver Spring—media-savvy, ruthless and uncompromising." The new groups formed in the 1980s differed from earlier animal protection organizations in at least two important respects. First, they were staffed largely by grassroots activists, young people eager to march and protest, not salaried careerists with families to feed and mortgages to pay. Second, they were in many cases inspired by the work of radical philosophers like Peter Singer and Tom Regan and claimed to base their objections to the human use of animals on rational argument, not emotional appeals.

The new activists disdained the moderate style and modest goals of the humane movement. Instead, they adopted the militaristic rhetoric and no-compromise platform of the nineteenth-century antivivisection movement, while greatly expanding the scope of its critique. Not only was animal research a morally corrupt practice, but so were meat eating, the manufacture and consumption of fur and leather, and the imprisonment of animals in zoos, circuses, and aquariums. "The animal rights view holds that human utilization of nonhuman animals, whether in the laboratory, on the farm, or in the wild, is wrong in principle and should be abolished in practice," animal rights philosopher Tom Regan wrote. "Whatever humans might gain from such utilization (in the form of money or convenience, gustatory delights, or the advancement of knowledge, for example) are and must be ill-gotten."

Proponents of animal rights assert that favoring the interests of the human species over those of other animals (speciesism) is akin to favoring the interests of one's own race (racism) or gender (sexism). Each of these "isms," one no less than the other, is an equally malignant and malicious tendency that must be eradicated to facilitate the birth of a new earth. As Jasper and Nelkin pointed out in their 1992 study of the movement, "Animal rights is a moral crusade. Its adherents act upon explicit moral beliefs and values to pursue a social order consistent with their principles. Their fervent moral vision crowds out other concerns." Jasper and Nelkin noted that like other moral crusaders (for example, antiabortionists, antipornography proponents), animal rights campaigners are "moral missionaries" who "often insist they have no other broader partisan agenda. They are less interested in material benefits for themselves than in correcting perceived injustices. Animals are a perfect cause for such a crusade: seen as innocent victims whose mistreatment demands immediate redress, they are an appealing lightning rod for moral concerns."

Jasper and Nelkin defined three types of animal protectionists: welfarists, pragmatists, and fundamentalists. Welfarists, the traditionalists of the movement, "accept most current uses of animals, but seek to minimize their suffering," while both pragmatists and fundamentalists are committed to fundamental change in human-animal relations. However, while pragmatists are willing to work to "reduce animal use through legal actions, political protests, and negotiation," fundamentalists "demand the immediate abolition of all exploitation of animals, on the grounds that animals have inherent, inviolable rights." The latter group often compares human use of animals to the enslavement of blacks, despite the problematic associations inherent in such a comparison. Nonetheless, "although far less numerous than pragmatist or welfarist organizations, these groups set the tone of the new animal rights movement," Jasper and Nelkin said.

Without question, the most successful of the new breed of animal rights groups to appear in the 1980s was PETA, and PETA grew rapidly. "By the early 1990's, PETA claimed more than 400,000 members, a paid staff of over 100, an annual budget of nearly $10 million," Deborah Blum wrote in *The Mon-*

key Wars. In the early years, PETA seemed to possess both an unerring sense of public sensitivity to the animal issue and a talent for choosing the kind of targets and battles that would receive extensive media coverage. However, the organization made enemies early on, not only among its natural adversaries, but within the movement as well. The kind of heavy-handedness described by Jo Shoesmith raised hackles among other activists. Henry Spira's distaste for PETA's leadership was well known although not publicly advertised, in the interests of movement unity. However, the rift became public in 1987, when Spira defended Procter & Gamble during a PETA attack, noting that the company was one of the leaders in the growing movement within industry to develop alternatives to animal testing. "It seems to me that when a corporation is responsive to our concerns, it makes no sense to clobber them over the head. Rather, we want to encourage them to continue to be responsive and use their responsiveness as an example to others," he said. Spira, whose Animal Rights International was essentially a one-man operation (albeit with thousands of supporters), was also critical of PETA's large staff and its use of the type of direct-mail fund-raising necessary to support such an organization.

"Despite its success, or perhaps because of it, PETA is controversial," Jasper and Nelkin commented in 1992. "Other animal protectionists have accused it of 'taking over' the New England Anti-Vivisection Society by packing a meeting with its own supporters, many flown in from Washington. In 1987, members of the animal protection community received an anonymous letter denouncing PETA for 'no elections, closed decision-making, and rigid hierarchy.' It claimed 'Newkirk will not tolerate any actions or thoughts which challenge her power over others,' and said 'PETA is on a Sherman's march through the animal liberation movement, a warpath where anything goes if it's 'for the animals, in their view.' "

In a 1985 interview conducted by Charles W. Griswold, Jr., an associate professor of philosophy at Howard University in Washington, D.C., Ingrid Newkirk defended her organization and its tactics. Commenting at the start of the interview that PETA had "close to 90,000 members now, and we're growing by leaps and bounds," Newkirk said that her organization "brought animal rights out of the closet across the country, showing people things they can do in their own lives. Suddenly people felt that they could put their feelings into action." Questioned by Griswold about the 90,000 supporters, she admitted that only 60 to 70 percent of that number were dues-paying members of the organization. "We don't require that members pay dues, just that they are doing something for animals."

Griswold next asked about the relationship between PETA and the Animal Liberation Front—"We speak for them," Newkirk replied—and whether or not she agreed with the destruction of property that was becoming a hallmark of the group's activities. "I don't know whether one agrees or disagrees with those tactics. We don't choose to do those tactics here. I certainly more firmly disagree with the people who destroy and mutilate living beings. The real violence is on the part of those who use animals for their own gain," she said. Contending that "there's a holocaust for animals going on in our land," she added that "to have people not take action like that wouldn't make sense."

Griswold then questioned Newkirk on her comparison between animal rights and civil rights. "You've compared several times the way in which we treat animals with the way people used to treat blacks in the United States and elsewhere," he said. "You think there's a close analogy between what in the literature gets called 'speciesism' and things like racism . . . you view this as not just cruel but as wrong in the same way that it was wrong for whites to enslave blacks or Hitler to kill the Jews." "Historically," Newkin replied, "it is the same mind-set, it is the same prejudices, it is the same disregard for the rights or interests of other races, gender, what have you, that have resulted in great massacres."

Griswold grilled Newkirk on the philosophical and practical ramifications of this perspective, articulated by the PETA founder as opposition to "the use of animals in any pursuit where it's the oppression of one species over others. The use of any individual by other individuals . . . we are absolutely opposed to one group deciding that another group exists for its use." He was particularly persistent in his questions about the impact of animal rights philosophy on biomedical research. In response to his questions, Newkirk denied that animal research had produced significant benefits with respect to human health and introduced a theme prominent in nineteenth-century antivivisection literature, the negative effects of vaccination, with a contemporary twist. "I believe that we're now beginning to learn—and many people are stopping to vaccinate their children—that we have wreaked havoc on our immune system by such a dependency on animal-based vaccines in our youth. We have injected animal proteins into our bodies that have lain dormant for years and are now crossing the species barrier and doing all sorts of things."

Throughout the interview, Newkirk labeled the belief that research on animals had helped eradicate infectious diseases like polio "a misconception." When asked by Griswold, "If it could be shown to you by a surgeon you respect that an experiment on an animal was in fact indispensable for some procedure, say surgery, to save your life, would you still reject it?" she answered, "You have no right. The thing is that the question is not real. The situation does not exist. We shouldn't have the choice. We shouldn't be able to judge whether my neighbor's child should die in order to have my child live."

"In other words, the answer is no," Griswold said.

"If you want a yes-no interview, you'll have to go somewhere else, won't you?" Newkirk replied.

PENNSYLVANIA, 1984

IN THE VIDEOTAPE an attractive young woman holds up a brain-damaged monkey. Off-camera, a male voice is heard saying, "You'd better hope that the antivivisection people don't get ahold of this film." "Who?" the woman says. A few moments later, after more off-camera joking and high jinks, the monkey looks up at the woman imploringly, and the male voice speaks for him: "You're going to rescue me from this, aren't you? Aren't you?"

This is the final scene in a twenty-minute videotape called "Unnecessary Fuss," distributed by PETA in June 1984. There was no way that researchers could argue that this scene at the University of Pennsylvania Head Injury Clinic had been staged (as some had alleged of the photographs taken by Peter Hamilton and Alex Pacheco at Edward Taub's Institute for Behavioral Research). The PETA videotape had been created using seventy hours of videotape made by the scientists at the U.P. clinic themselves, documenting their work on head injury experiments funded by the National Institutes of Health since the early 1970s. Thomas Gennarelli and Thomas Langfitt, researchers at the university, had received about $330,000 from NIH in 1984 for this work, which sought to create injuries, similar to those received by human beings in auto accidents and other traumas in which the brain is damaged, by rapid slamming against the skull. In the tape, this process is called banging by one of the researchers—as in "This monkey has been banged twice."

The tapes were stolen from the head injury lab during a Memorial Day weekend raid by the Animal Liberation Front. Alex Pacheco edited the tapes, and Ingrid Newkirk provided narration for scenes of researchers smoking and eating while performing experiments, using screwdrivers and hammers to knock off the helmets of recently injured animals (in one case taking off part of an animal's ear), performing surgery in conditions far from sterile, and mocking the injured animals. These graphic scenes were interspersed with repeated shots of "banging" heavily restrained animals that appeared to be conscious and quotes from university officials and research scientists involved in the experiments denying that the animals were mistreated in any way (for example, "We treat the baboons like human beings" and "The university has nothing to apologize for"). The combination of graphic maltreatment, in clear violation of Public Health Service policy, the jocularity and insensitivity exhibited by the researchers, and the denial of responsibility by university officials and the principal investigators was deeply disturbing. The ALF and PETA could not have manufactured a more damning indictment of the way research was being conducted in some laboratories, hidden from Department of Agriculture inspectors, members of Congress, and the public.

Segments of the tape were shown on national television and predictably resulted in a public relations nightmare for the university. However, the research community quickly closed ranks and defended the work. NIH instituted a year-long investigation of the laboratory that resulted in a renewal of the grant for five more years. In response, PETA organized a sit-in at NIH headquarters in Bethesda, Maryland, in July 1985. After four days of nonviolent protest and increasing media attention, Secretary of Health and Human Services Margaret Heckler suspended the grant "until all questions about the use of primates in these head injury experiments have been satisfactorily resolved." One month later the university was fined $4,000 for violations of the Animal Welfare Act, and soon after, it announced that it was suspending all primate research in the Head Injury Clinic. Once again, the protesters had won.

Over the next few years, the ALF continued raiding various research facilities, with mixed results. In a few cases, abuses were uncovered and funding was temporarily suspended until the institution corrected the problems. In other cases the ALF simply "liberated" animals, destroyed files, and vandalized laboratories. The Association of American Medical Colleges estimated "the cost of lost data, break-in damage, property defacement, and demonstrations" at more than $3.5 million from 1985 to 1990. Individual researchers targeted by the ALF faced not only the loss of equipment and data, but were also subject to personal harassment. John Orem, the sleep studies researcher quoted in the introduction to this book, lost over $50,000 worth of equipment in an ALF raid on July 4, 1989, and received death threats after the incident, according to an article published in *the Chronicle of Higher Education*.

"The ALF was made up of very ordinary people with very ordinary jobs," Jo Shoesmith said in our 1998 interview, choosing her words carefully. "They fit a certain demographic profile that made infiltration easy. Very well dressed, very well spoken, smooth-talking. They were a hard population to trace, to track down, because they were everything that you did not expect, kind of like 'the little old lady in tennis shoes.' " Shoesmith said that the property-damage phase of ALF activity was a late development. "The rationale being circulated in the movement was 'We're just doing economic damage to the industry and creating a fear strategy.' People believed that it was having an effect."

Indeed, universities invested large sums of money in security systems, and scientific spokespeople began to talk about a fortress mentality. Researchers felt besieged, and in 1992, the Animal Liberation Front was listed as one of the ten most dangerous terrorist organizations in the United States by the FBI. As the economic and psychological costs of a sustained campaign of harassment escalated, the National Association for Biomedical Research and other research defense organizations began to lobby Congress for legislation making acts such as those that resulted in the "Unnecessary Fuss" video federal crimes. These efforts achieved success when the Farm Animal and Research Facilities Protection Act was signed into law on August 26, 1992.

Although many individuals and organizations working in animal welfare and animal rights (including Christine Stevens, Henry Spira, Peter Singer, and other leaders of the movement) publicly condemned violence, by the end of the decade an initially sympathetic public began to associate animal rights with terrorism. The actions of the ALF and the unwillingness of PETA spokespeople to condemn such behavior were probably partly responsible for this shift in public sentiment, but the research community was also quick to seize on the opportunity to associate all activism on behalf of animals with the type of radical who was willing to burn down laboratories and harm those who disagreed on the need for "empty cages." As the Animal Liberation

Front became more violent and destructive, and the scientific community responded with its own verbal artillery, the conflict between science and animal protection became more volatile and more polarized than it had been since the early years of the century. It became an all-out war—and every war has its casualties.

SEVEN

STALKING THE SHADOW

While some traits peculiar to the shadow can be recognized without too much difficulty as one's personal qualities, in this case both insight and good will are unavailing because the cause of the emotion appears to lie, beyond all possibility of doubt, in the other person. . . . Projections change the world into the replica of one's unknown face.

—Carl Jung,
*Aion: Researches into the
Phenomenology of the Self*

QUEENS, NEW YORK, SEPTEMBER 1988

A HANDSOME, dark-haired man approaches a solemn, suspicious woman in a pizza parlor and strikes up a conversation. He tells her that his German shepherd has just given birth to a litter of puppies. The woman's demeanor changes utterly. As she begins to tell him about her own four dogs, whom she describes as "my babies," her face begins to glow and she looks almost pretty. He notes this later as he prepares a report for his employers, who are paying him nearly $500 a week to befriend this dog-loving woman. The man's name is Marcus Mead, and he is a petty criminal; he has served time for bouncing checks and for mail fraud. The woman, thirty-two-year-old Fran Stephanie Trutt, has never committed a crime, but Mead's employer, a corporate security firm called Perceptions International, has had her under 24-hour surveillance for months.

Another agent of the company, Mary Lou Sapone, has already

befriended the woman and discovered that she is a near recluse and appears to have no friends but her dogs. She is, however, filled with anger at her desertion by a lover who Sapone initially believes is a man but who Trutt later confesses is a fifty-seven-year-old Colombian woman. As Sapone's tape recorder rolls, Trutt shares intimate details of her relationship with this woman and pours forth her rage and resentment at her lover's rejection. Trutt tells Sapone that she would like to kill her. "Just like Leon," Sapone prods, bringing the conversation around to the area of interest of her employer. "Yeah, him too," Trutt mumbles.

Sapone is pleased. Since June 1987, she has been attending animal rights meetings and marches around the country, filing reports with her employer, and chatting up men and women in the movement. She's been prodding and probing, suggesting violent acts to see just how far these folks are willing to go in their desire to protect animals. Most of the people she's met shrug off her suggestions that they bomb researchers' homes or cars.

But in April 1988, she'd had a stroke of luck. She'd met Fran Stephanie Trutt at a demonstration in Norwalk, Connecticut.

Since 1981, animal rights activists had been holding regular demonstrations in Norwalk, at the headquarters of the United States Surgical Corporation. U.S. Surgical is an enormously successful manufacturer of biomedical devices and tools—in 1987, the company boasted $252 million in sales. U.S. Surgical was reported to control 70 percent of the world's lucrative surgical-staple market at that time. The company used dogs (over one thousand a year) in training salespeople who demonstrated the use of the product to surgeons. The animals were destroyed after each session, and the corporation's supplier, Quaker Kennels, had been cited for repeated violations of the Animal Welfare Act. Activists insisted that there were other, better ways to demonstrate the product and that the training sessions with dogs were just a slick marketing ploy.

Trutt, who lived in New York City, caught rides with fellow activists to attend the demonstrations and had been videotaped by the company with the other protesters, marching around the parking lot, holding placards, and shouting "Staple Hirsch." Leon Hirsch was U.S. Surgical's founder and its chairman of the board. He did not like protesters in his parking lot, and he believed that the growing power of the animal rights movement was a threat to his business. The public was increasingly restive, questioning the need for things like cosmetics testing on rabbits and product training sessions using dogs. That's why Hirsch had hired a corporate security group, Perceptions International, to spy on the movement for the past several years. Marcus Mead and Mary Lou Sapone worked for Perceptions International. And according to Mary Lou Sapone, Fran Trutt hated Leon Hirsch and wanted him dead.

Of course, lots of people in the animal rights movement hated Hirsch and wanted him dead. They screamed out their hatred in his parking lot. But with a little encouragement, Fran Trutt seemed capable of attempting to carry out the deed. Research people had been saying for a long time that the next step for the animal rights crowd was personal violence—an attack on a scientist. Fran Trutt appeared to be that next step—with a little help from Marcus Mead and Mary Lou Sapone.

NORWALK, CONNECTICUT, NOVEMBER 10, 1988

IT WAS DARK when Mead and Trutt pulled into the parking lot of U.S. Surgical corporate headquarters, just before midnight. Mead was nervous, even though he had rehearsed this scenario earlier that day with Sgt. Thomas Fedele of the Norwalk police and U.S. Surgical's chief of security, James Rancourt. Step-by-step they had walked Mead through the procedure, telling him exactly where to park and what to do when they appeared. "The police wanted us to be in the fenced area so that when they came after us, she would have no way of escaping," Mead later told reporters. Still, he was nervous, sweating bullets. After all, he had just driven from Queens, New York, to Norwalk, Connecticut, with a bomb in his car.

On Tuesday, November 8, Trutt had informed Mead that she had some bombs. "Call me on Thursday and let's decide what to do then," Mead told her. The next morning he called his boss, Jan Reber, head of Perceptions International. Reber told Mead to bring Trutt and her bombs to U.S. Surgical headquarters on Thursday night, November 10. "She called me on Thursday," Mead continued, describing the chain of events to the *Westport News*, a local paper, "and I said, 'Why don't we do it tonight?' She said, 'Tonight?' and I said, 'Yeah, tonight.' And she said, 'Well, okay.' I didn't have to talk her into it."

On the way to Connecticut, Trutt called Sapone. "She seemed nervous about what she was doing and mad at me for not being available," Sapone reported to her bosses. Trutt later told reporters that she was reluctant to go through with the bombing on November 10 and that she had called Sapone from a highway rest area on the way to Connecticut. Sapone convinced her to proceed, telling her to trust Mead. "He knows what he is doing," Sapone said.

The plan worked flawlessly. Trutt and Mead drove up in his truck before midnight. Trutt placed the bomb in bushes near Hirsch's parking space. As she and Mead were walking back to his truck, police and security officers jumped out from another clump of bushes, shouted "Freeze!" and pulled out guns. Mead jumped over a fence and crossed the street. No one followed him. Trutt was forced to the ground, handcuffed, and taken to police headquarters to be booked for the attempted murder of Leon Hirsch, chairman of the board of U.S. Surgical. The next day it was all over the papers. The fears of the research community were well founded, it seemed. The animal rights nuts had gone from demonstrations to lab break-ins to attempted murder. They were crazy. Researchers feared for their lives. Commenting on the attack at a press conference, Leon Hirsch said, "We are not going to be intimidated by murderers and terrorists who are bent on abandoning mankind in favor of animals. If you stop research today, you give up hope of a cure for AIDS and other diseases in the future."

Sapone and Mead congratulated themselves on a job well done. There was only one problem. When Mead, who considered himself a hero, began to boast about his role in the arrest of Trutt, Perceptions International denied that he worked for them. Worse, after he gave an interview describing his involvement to the *Westport News* in January, in which he unwittingly bolstered Trutt's claim that she had never intended to kill Hirsch, only to frighten him, he was arrested by the Norwalk police on a two-year-old warrant for passing a bad check. "I never heard her say that she wanted to kill Hirsch. All she said was that she wanted to commit some sort of fanatical act against him," Mead said. But everyone was acting like he was some kind of criminal, like he was Trutt's accomplice. "I thought Perceptions International had my interests at heart, but they won't even admit they know me. After two months I'm running across people who are questioning my role, and that's nagging at me," Mead told *Westport News* reporter John Capsis. "I knew who the client was from the beginning," he added. "The checks came from Perceptions International, but I was told I was being paid by U.S. Surgical."

"U.S. Surgical in no way encouraged, orchestrated, financed, or in any way facilitated Ms. Trutt's activities," Leon Hirsch countered in a prepared statement during a press conference at U.S. Surgical's corporate headquarters on January 13, 1989. He left in anger as reporters continued to question him about U.S. Surgical's connections with Perceptions International. Meanwhile, Perceptions arranged for lawyer Kenneth D'Amato to represent Mead and covered his legal fees. D'Amato was a member of the board of Perceptions International. He barred his client from speaking to the media.

Trutt, meanwhile, was sitting in the women's prison in Niantic, Connecticut. She was frantic with worry about her dogs. Her friend, Mary Lou Sapone, agreed to keep the dogs for her while she was incarcerated. Trutt knew she had been set up. She complained about Mead to Sapone. Every conversation was recorded by Sapone and submitted to her superiors at Perceptions. It wasn't until months later, when Sapone's role in the plot was broken in the Norwalk papers, that Trutt confronted her erstwhile "friend." "I hate this when you believe them over me," Sapone said.

Trutt was in jail for nearly two years before the case came to trial. Her public defender had been replaced by a civil rights attorney, John Williams. Williams established a paper trail linking Mead and Sapone to Perceptions, and Perceptions to the U.S. Surgical Corporation. Perceptions International had paid the two agents nearly $90,000 and had hired a Long Island firm called Hallmark International to conduct 24-hour surveillance of Trutt in the months before the attempted bombing. "Hirsch clearly must have known about the purchase of the bombs as early as last summer—long before she had the chance to use them," Williams told journalists. He also charged "that there was a sufficiently intimate relationship between U.S. Surgical and its people and the Norwalk police, including Fedele, and then later with the prosecutor as well that they can't escape the taint of what the corporation did." Norwalk police often worked as security at U.S. Surgical in their off-hours.

His client had been set up, Williams was to argue in court. It was a clear-cut case of entrapment, and his client was eager to go to court and clear her name. Meanwhile, prosecutors announced their possession of Sapone's tapes, in which Trutt spoke of her desire to kill Leon Hirsch and described in intimate detail her relationship with her much older female lover. Listening to the tapes during a recess in the pretrial hearing, Trutt wept. Her attorney said that she was humiliated by the idea of her sexuality going public. She was also devastated by her separation from her animals, particularly after a federal judge rejected her request to see her dogs while incarcerated. Prosecutors offered Fran Trutt a bargain: plead no contest to the charge of attempted murder and spend thirty-two months in jail (minus time served) with three years' probation as part of a ten-year suspended sentence. Williams urged her to reject the plea. He wanted to put Leon Hirsch and U.S. Surgical on trial. Hirsch was not particularly popular, having built what neighbors described as "an armed compound," complete with six-foot electric fence, armed guards, and trained attack dogs, on 132 acres of land he owned in New Canaan. But after hearing the tapes and realizing that she would have to spend little more than a year in jail if she accepted the sentence, Trutt said no.

Because the case never went to trial, evidence linking U.S. Surgical and its agents at Perceptions International to the crime was never heard in court. Although the case was heavily covered in local papers—the *Hour*, the *Advocate*, and the *Westport News*— by reporters John Capsis, Margaret Tierney, Mike McIntire, and Beth Cooney, the only substantive national coverage was a *New York Times* article by Celeste Bohlen, "Animal-Rights Case: Terror or Entrapment," on March 3, 1989. Despite the fact that the Connecticut reporters obtained transcripts of telephone conversations and copies of reports by the undercover agents to Perceptions International and established a paper trail leading to U.S. Surgical, the Trutt case was used to bolster contentions that the animal rights movement was out of control and that researchers had much to fear from activists. A report by the Maldon Institute, a conservative think tank, dated April 12, 1991, and titled "Animal Rights: Militancy and Terrorism," profiles the origins and activities of the movement and ends with a summary of the "U.S. Surgical Action" without any acknowledgment or discussion of the problematic chain of events that led Fran Trutt to the U.S. Surgical parking lot on the night of November 10, 1988.

The level of paranoia within the U.S. research community increased exponentially after the Trutt incident. The Silver Spring monkey case and the University of Pennsylvania Head Injury Clinic break-in had created public outrage and a demand for stricter regulation of biomedical research. The research establishment, forced to respond to public interrogation and animal rights pressure, by 1988 had become increasingly resistant to further concessions. Fran Stephanie Trutt appeared just when hard-liners most needed her. She was portrayed as the type of anonymous mentally disturbed individual who broke into labs and left hate-filled messages for researchers. She was a researcher's worst nightmare come to life. Any one of us is personally vulnerable, scientists felt. Any one of us could be next. Without a trial and without substantial media coverage, the role played by Perceptions International agents, acting for U.S. Surgical, was ignored and finally forgotten.

BY 1988, after nearly a decade of sustained aggressive animal rights activism, changes were taking place in the practice of research in the United States. Animal experimentation had not been abolished, but it was regulated more strictly each time the 1966 Laboratory Animal Welfare Act was amended (1970, 1976, and 1985). Although the original law applied only to cats, dogs, primates, rabbits, hamsters, and guinea pigs on dealers' premises and in laboratories prior to experimental use, coverage was eventually extended to all warm-blooded animals (except rats, mice, and birds), with protection encompassing the duration of their lives in laboratories. The 1985 amendments (Improved Standards for Laboratory Animals Act), passed in the wake of the University of Pennsylvania Head Injury Clinic debacle, established an information service in the National Agricultural Library in Beltsville, Maryland. This Animal Welfare Information Center (AWIC), was created by Congress to meet the provisions of the amendment, which required investigators to consider alternatives to experiments that might cause pain and distress. The AWIC, working in cooperation with the National Library of Medicine, enabled researchers to find alternatives to painful experiments. The National Association for Biomedical Research actively opposed the creation of the AWIC, arguing that alternatives were a chimera devised by antivivisectionists to make the public believe that experimental research on animals was no longer necessary.

The 1985 amendments also required registered research facilities to establish Institutional Animal Care and Use Committees (IACUCs) to review all experimental protocols involving animals. IACUCs were modeled on the Institutional Review Boards

(IRBs) set up in research institutions to protect human subjects in the wake of public outrage following the revelation of the Tuskegee Syphilis Study. According to the legislation, IACUCs were to include at least five members, including a veterinarian and an "unaffiliated" member representing community interests in the humane care and treatment of research animals. IACUCs were charged with reviewing every experimental protocol submitted by scientists in research institutions. The IACUC review was to address animal welfare concerns, not to establish the scientific merit of experiments.

These reforms were greeted with enthusiasm neither by many scientists nor by organized research interests. For many, they constituted an attack on academic freedom and represented bureaucratic meddling in purely scientific affairs. IACUC review was perceived as yet another hurdle that investigators would be forced to jump in the increasingly arduous process of applying for grants. The resistance of the U.S. research community to examining its use of experimental animals was somewhat perverse, as the number of animal experiments had been dropping steadily since the late 1970s in both the United States and Europe.

In England, where detailed statistics were compiled by the government, the trend was particularly well documented and illustrated a sharp rise in animal experiments after World War II that peaked in the mid-seventies, with a steady decline thereafter. Andrew N. Rowan, currently senior vice president for research, education, and international programs at the Humane Society of the United States, notes that although the U.S. numbers are less reliable than those of Great Britain because of methodological problems, "it nonetheless appears as though animal use (or at least the use of the six species counted by USDA) has declined by at least 23% and maybe as much as 40% since 1967."

Rowan says the evidence suggests that the steady decline in animal use has continued, and perhaps more rapidly since 1980, as industry has reduced the numbers of animals used in testing protocols. "Hoffman-LaRoche reported that its use of animals dropped from one million a year to 300,000 during the 1980's even though the number of new drug entities under investigation remained about the same," he wrote in *The Animal Research Controversy: Protest, Process and Public Policy*, published in 1995.

In a 1998 interview with the author, Mark Matfield, director of the Research Defense Society in Great Britain, attributed the decline in animal experiments to a number of factors, including "economics (the cost of laboratory animals increased dramatically), more rigorous implementation of regulations, shifts in the direction of science away from whole animal work to in vitro studies, and growing public concern and protests about animal experimentation." In the United States, these factors were equally significant. However, massive changes in product safety testing practices also had a major impact on the numbers of animals used by industry, and these changes in testing practices compelled a reexamination of the manner in which animals were used in research as well. Although the academic research community had traditionally fought hard against increasing public pressure for reform, industry reacted differently when consumers began to demand "cruelty-free products" in the early 1980s. Once again, Henry Spira played an instrumental part in this story.

After the passage of the Food, Drug and Cosmetic Act in 1938, manufacturers were responsible for insuring the safety of these products before placing them on the market. This law was passed following a series of disastrous occurrences in the 1920s in which consumers were seriously injured by untested cosmetics. The push for legislation was given greater impetus in 1937, when over one hundred Americans died by poisoning after sulfanilimide, an early antibiotic, was mistakenly mixed with a toxic solvent and marketed as Elixir of Sulfanilamide. In 1958, the Delaney Amendments to the FDC Act required that manufacturers also provide data proving that their products would not cause cancer. By the late 1970s, at least four government agencies, including the Food and Drug Administration, the Environmental Protection Agency, the Consumer Product Safety Commission, and the Occupational Safety and Health Administration, required safety testing on the various chemicals and ingredients used in a wide variety of consumer products—from mascara and tooth-paste to lawn chemicals and pesticides. A number of other government entities either required or conducted testing as well.

One of the basic tests, for eye irritation, carried out on nearly all products in both government and corporate laboratories was named for a Food and Drug Administration scientist who standardized an existing protocol in 1944. The Draize test for eye irritation was not pretty. In a typical Draize test, a test substance was placed in one eye of a restrained rabbit, and scientists observed changes in the cornea, conjuctiva (the pink fleshy membrane around the eyeball), and iris. The rabbit's eyes were inspected over a seven-day period to determine whether damage to the eye was temporary or permanent. The Draize test was standard and was required by regulatory authorities throughout the world. It wouldn't be challenged until 1980, when, following the successful campaign at the Museum of Natural History and the repeal of the Metcalf-Hatch Act in New York State, Henry Spira and his new science adviser, a South African biochemist named Andrew Rowan, strategized about their next course of action.

Rowan had spent two and a half years following his graduation from Oxford working for a group in England called the Fund for the Replacement of Medical Experiments (FRAME). FRAME was founded in London in 1969 by Dorothy Hegarty, a visionary but rather imperious woman who reportedly clashed with Muriel, Lady Dowding, another strong-willed antivivisectionist, when both were active in the National Antivivisection Society. Having left that organization and started her own group, Hegarty, an ardent vegetarian, was influenced by two scientists—her son, Terence Hegarty, and Charles Foister—to reject the abolitionist position advocated by the antivivisection societies and to steer a middle course between abolition and the research community's commitment to continued high levels of animal use. FRAME's moderation succeeded where others had failed, and by 1984, the group was working with the government to revise the 1876 Cruelty to Animals Act and to implement new legislation in the United Kingdom and the European Union.

While working at FRAME, Andrew Rowan became familiar with a 1978 book called *Alternatives to Animal Experiments* by British scientist D. H. Smyth. Smyth, a physiologist and the chair of the Research Defense Society in the United Kingdom, suggested that because of the increasing sophistication of science, as well as animal welfare concerns, the Draize test was ripe for replacement. With characteristic pragmatism, Spira, advised by Rowan, realized that the most vulnerable target in a campaign against the Draize test would be the cosmetics industry. Women had traditionally exhibited the most sensitivity to the suffering of experimental animals—the nineteenth-century antivivisection societies had been heavily female—and no doubt modern women would be appalled if they began to associate the quest for beauty and the myriad cosmetic products they used every day with animal cruelty. Spira targeted an industry leader—the cosmetics giant Revlon.

On April 15, 1980, a full-page ad sponsored by Spira's Coalition to Abolish the Draize Test appeared in *The New York Times*. HOW MANY RABBITS HAS REVLON BLINDED FOR BEAUTY'S SAKE? the headline inquired, superimposed over a picture of a cuddly rabbit about to receive an eyeful of an unidentified liquid. (This photograph, like all of the ads used in Spira's subsequent campaigns, was created by the Coalition for Animal Rights' public relations adviser Mark Graham.) The Revlon campaign proceeded along the same lines as the Museum of Natural History protest, with the company's initial resistance gradually melting away as public relations pressure increased. Within a year the company had donated $750,000 to the Rockfeller Institute to fund research into alternatives to the Draize.

Spira and associates next sent a letter to Avon, suggesting that the company review its testing procedures. Avon's hierarchy, not wishing to repeat the Revlon experience, contacted their industry trade group, the Cosmetics, Toiletry and Fragrance Association

(CTFA), headquartered in Washington, D.C. CTFA represents not only manufacturers of products but also the suppliers who produce the ingredients used to create the vast array of cosmetics, perfumes, soaps, deodorants, mouthwashes, and other personal care and hygiene products so much a part of modern life. The cosmetics trade group, fully aware of the industry's vulnerability in the face of animal rights activists' claims that animals were being sacrificed and forced to suffer so that companies could develop a new kind of mascara or mouthwash, established a $1 million fund to support the development of alternative tests. Universities were invited to submit applications to administer the grant money. D. H. Henderson, a physician who had headed the World Health Organization's campaign for the global eradication of smallpox, heard about the grant money when he found himself sitting next to the CEO of Noxell Corporation, a Baltimore-based cosmetics company (maker of Noxema) on a flight back to the city. On his return to Baltimore, Henderson, who was at that time dean of the Johns Hopkins School of Public Health, discussed the grant with neuroscience researcher Alan M. Goldberg, who had developed an in vitro test for neurotoxins, and encouraged him to apply.

In September 1981, the CTFA and the university announced the founding of the Johns Hopkins Center for Alternatives to Animal Testing (CAAT), based in the School of Public Health. In its first year, the center held a scientific meeting in Baltimore where many of the people who were to play an instrumental role in the search for alternatives met for the first time. Within five years, CAAT's list of corporate sponsors read like a Who's Who of American industry—3M Corporation, Exxon, Hoffman-LaRoche, IBM, Johnson & Johnson, Procter & Gamble, Bristol-Myers Squibb. By then it had become clear that not only manufacturers of cosmetics were going to be targeted. Any company using animals for testing was vulnerable.

Initial reaction to the forming of this center within both the animal rights and scientific communities was outright cynicism. It was a public relations ploy, animal rights activists were convinced, meant to take pressure off companies who were under attack not only by Spira and his allies but by other, more radical, animal rights campaigners who demanded not a leisurely search for "alternatives," but immediate abolition of the Draize and other animal tests. Scientists, on the other hand, aware of the international regulatory apparatus mandating testing of products and the legal and moral obligation of companies to establish product safety—a mandate carried out entirely by animal testing for more than forty years—were not sanguine about the prospect of change. Surprisingly, change did come, more rapidly than most people had anticipated, and it was not based solely on animal welfare criteria, although the vulnerability of industry to animal rights protests certainly provided the necessary impetus to get the ball rolling.

By the mid-seventies, animal testing had become an enormously expensive and cumbersome process. Unlike research, which is an open-ended quest to discover previously unknown facts, testing utilizes internationally accepted protocols to establish particular outcomes. Although to many members of the public, the phrase "animal testing" implies the entire range of experimental uses of animals, to scientists it indicates only one category of animal study, those intended to establish the safety or efficacy of various products, whether industrial chemicals, drugs, or cosmetics. Tests carried out in Japan to establish the safety of a pesticide are much the same as those used in Germany or in the United States, because every nation works according to the same standard set of guidelines—and, in 1981, most of them were based on the use of whole animals.

In the United States alone, a number of government agencies either regulate or conduct testing of a wide range of products, based on legislative mandates. Aside from the Food, Drug and Cosmetic Act already mentioned, which assigns regulatory power to the Food and Drug Administration, the Toxic Substances Control Act of 1976 gives the Environmental Protection Agency power to ban or restrict the manufacture or use of potentially hazardous chemicals. The Act also authorizes the EPA to require testing of potentially hazardous chemicals already on the market. Similarly, the Occupational Safety and Health Administration, responsible for monitoring workplace safety, uses toxicological testing to make regulatory decisions. The Department of Transportation oversees labeling of potentially harmful substances transported across state lines, and the Consumer Product Safety Commission serves as a clearinghouse for information on the potential hazards of a broad range of consumer products. Individual companies are also responsible for ensuring the safety of their products via in-house or contract testing. The same is true of their suppliers, who manufacture the raw materials for a broad range of products, from shampoos to pesticides.

For decades, the hazards or safety of drugs, chemicals, consumer products, and their ingredients have been evaluated by batteries of whole animal (in vivo) tests that measure the effects of limited exposure of animal subjects to high doses of a substance (acute toxicity) and repeated long-term exposure (chronic exposure), as well as specific end points such as cytotoxicity (ability to damage cells), mutagencity (ability to cause changes in genetic material), carcinogenicity (ability to cause cancer) and teratogenicity (ability to cause birth defects). All these tests are used to determine the potential hazards of a substance and safe levels of exposure. Together they form the experimental aspect of the science of risk assessment, the process by which substances are evaluated for their potential impact on human health and safety. However, given the enormous numbers of products and chemicals covered by existing legislation, most of which have never been fully tested, our knowledge of the effects of these chemicals on human health and the environment remains limited. As late as 1997, representatives of the Environmental Defense Fund charged that the health effects of more than 70 percent of chemicals produced in high volume remain unknown.

In the early eighties, it was clear to many people in industry that the development of in vitro tests capable of assessing potential hazards quickly and accurately would be a tremendous economic and scientific boon. Some scientists had been attempting to prove this argument for years, long before industry began to feel animal rights pressure. For example, Bruce Ames, a scientist at the University of California, introduced a cell-culture test for mutagenic chemicals in 1971, and by the mid-eighties the Ames assay was increasingly used in place of the much more expensive rodent carcinogenicity studies that had been standard since the passage of the Delaney Amendments in 1958. The Ames assay, which evaluated chemicals' mutagenic effects on the microorganism *Salmonella typhimurium*, wasn't perfect—not all carcinogenic chemicals are mutagens—but it saved a lot of animal lives and a lot of money by serving as a quick, efficient screening method capable of pointing out potentially hazardous chemicals. The same was true of many of the new in vitro tests being developed by scientists throughout the world in the wake of international public protest on the animal testing issue. Research groups such as ZEBET in Germany and MEIC in Sweden (both founded in 1989), ICAAT in the Netherlands and the European Center for the Validation of Alternative Methods (ECVAM) (both founded in 1992), have since provided funding for and a mechanism for review of new nonanimal tests.

These groups had their origins in the doubts expressed by a number of scientists about the efficacy of standard animal tests. For example, in 1981 Gerhard Zbinden, an internationally respected Swiss toxicologist, criticized one of the most basic tests in the toxicologist's arsenal, the LD50 (Lethal Dose 50). In the LD50, a group of animals was fed increasingly larger doses of chemicals to determine the approximate dose that would result in the death of half the test animals. Zbinden called this test "a ritual mass execution of animals" and questioned its usefulness as a measure of estimating the toxicity of chemicals to human beings. At the same time, other scientists were criticizing the validity of standard rodent carcinogenicity assays, in which animals were fed massive doses of chemicals (for example, saccharine) with regulatory policy based on the results. Scientists were beginning to question the assumptions on which these tests were based, and to search for a new testing strategy, one based on the new sciences of in vitro toxicology and molecular epidemiology. But because the development of these new strategies was more like "research" and less like "testing," there was a great deal of resistance in regulatory agencies and among toxicologists, who insisted that the old methods had inarguably worked to ensure safety, while the new ones remained, at best, untried. And besides, who was going to pay for all the research to develop new tests? Certainly not the government, which was having a difficult time keeping up with the sheer number of substances needing evaluation.

In the end, it was industry that funded the research that led to the development of new tests, because it was industry that was feeling the heat of consumer dissatisfaction. Customers were demanding products that were "cruelty free," a phrase that angered many scientists and regulatory officials, who were convinced that the real cruelty would be marketing unsafe products. Still, changes in testing practices were instituted—too fast for some people and not nearly fast enough for others. At the tenth-anniversary symposium of the Hopkins center, a CTFA representative was able to announce that the use of animals for eye irritation testing had been reduced by 87 percent since 1982 in the industry as a whole, with a 70 percent reduction in the number of animals used in other types of cosmetics testing. Meanwhile, individual companies announced similarly significant drops in animal use, with some announcing a moratorium on all animal use. By this time an increasingly broad array of products were being labeled NOT TESTED ON ANIMALS, although as many scientists were quick to point out, the phrase was deceptive because all of the ingredients used in those products had probably been tested on animals at some point over the past forty years, even if the finished product had not.

Nonetheless, a new paradigm was slowly taking shape whereby animal testing for certain uses (particularly cosmetics and personal care products) was increasingly viewed as a last resort, rather than a first. In the midst of these changes, a little-remembered book published in 1959 began to be discussed and referenced in scientific papers on animal welfare. Although certain people, notably Christine Stevens of the Animal Welfare Institute, had continued to promote the ideas advocated by Russell and Burch in *The Principles of Humane Experimental Technique* throughout the sixties and seventies, for the most part the book remained obscure until a new generation grasped its relevancy to the testing controversy.

Activists and scientists who discovered the book in the early eighties realized that it provided a blueprint for a new kind of collaboration and cooperation between animal protectionists and researchers. Early in the book, the authors provide a definition of the words *humane* and *inhumane* in the context of scientific studies: "Our use of these terms must not be taken to imply ethical criticism or even psychological description of persons practicing any given procedure. We assume throughout (probably with good grounds) that experimental biologists are only too happy to treat their animals as humanely as possible. The central problem then, is that of determining what is and what is not humane, and how humanity can be promoted without prejudice to scientific and medical aims. We must begin by examining the concept of humanity (or inhumanity) as an objective assessment of the effects of any procedure on the animal subject."

The book goes on to discuss questions of animal consciousness and the concepts of replacement, reduction, and refinement (the three R's) with respect to animal experimentation. Russell and Burch used experimental data to support their contention that in many instances it should be possible to either replace whole animals with cell- and tissue-culture studies, reduce the number of animals used, or refine the experiments (and the conditions in which animals are fed and housed) to reduce animal distress. "Replacement means the substitution for conscious living higher animals of insentient material. Reduction means reduction in the numbers of animals used to obtain information of given amount and precision. Refinement means any decrease in the incidence or severity of inhumane procedures applied to those animals which still have to be used."

Russell and Burch stressed that science and animal welfare were not incompatible goals; rather, good science depended on careful attention to animal well-being. "By now it is widely recognized that the humanest possible treatment of experimental animals, far from being an obstacle, is actually a prerequisite for successful animal experiment. Since the Second World War, in particular, this principle has been increasingly accepted," they commented in their introduction. "To approach this problem systematically is virtually to create a new discipline of applied science. Now that specializations are multiplying with unheard of rapidity, the creation of yet a new one may cause many hearts to sink; but this new science has the virtue of being a synthetic one, which brings together under a common view-point a vast variety of facts and ideas from a multitude of existing fields. Such synthetic disciplines are likely to be especially fruitful at the present stage of scientific evolution and from this one, apart from its immediate or long-term humanitarian fruits, we may expect an important contribution to the progress of the biological sciences at large."

The three R's quickly took hold among individuals and groups willing to accept some level of continued animal experimentation combined with gradual, if radical, reform. In 1990, Martin Stephens, vice president for laboratory animals at the Humane Society of the United States (HSUS), flew to England to meet W. M. S. Russell and Rex Burch. The HSUS was interested in instituting an annual award to recognize scientists who had done the most to advance the three R's during their careers, and it wanted to name the award after the two pioneers. Russell and Burch, who had not seen each other since the book was published, were flabbergasted and delighted. After decades of virtual silence, their ideas were finally receiving the reception they had hoped for when the book was published in 1959.

Discussing the impact of this book on the evolving debate, Andrew Rowan said, "Russell and Burch's book had little impact for the decade following its publication, but gradually it began to be referred to more and more often. By 1987 the industrial community had accepted alternatives, which began to be defined in terms of the 3R's. The academic research community was more resistant and to this day is unhappy about the term 'alternatives' and only marginally more satisfied with the idea of the 3R's."

Throughout the eighties and most of the nineties, the National Institutes of Health, with the exception of the National Institute for Environmental Health Sciences, strongly resisted the language and concept of alternatives, arguing that there are no true alternatives to the use of animals in research. Required by Congress in 1993 to "prepare a plan to conduct or support research which replaces, reduces, or refines the use of animals, establishes the validity of such methods, encourages their acceptance, and trains scientists in their use" (Section 205 of Public Law 103–43, the NIH Revitalization Act of 1993), the NIH one year later produced a *Plan for the Use of Animals in Research*, which reiterated its commitment to animal research while noting that certain grants and programs not specifically focused on the three R's nonetheless fulfilled the congressional mandate.

Louis Sibal, director of the Office of Laboratory Animal Research and chair of the committee charged with developing the plan, commented after its publication, "We lost a lot of time debating the word *alternatives*. Members of the committee didn't want to employ the word, because *alternatives* means *replacement* to the animal rights groups, and we wanted to talk about replacement, reduction, and refinement." Despite this purported embrace of the three R's within NIH, resistance to Russell and Burch's ideas in the U.S. academic community continues to this day and is perhaps best exemplified by the title of an article published by Herbert Lansdell, an NIH scientist, in 1993—"The Three Rs: A Restrictive and Refutable Rigamarole."

In 1989, Andrew Rowan, at that time an associate professor of environmental studies at Tufts University Veterinary School in Massachusetts, was sponsoring meetings between animal rights activists and scientists willing to talk to each other to find common ground. Attendees at these meetings included Adele Douglass of the American Humane Association, Henry Spira of Animal Rights International, Peter Theran and Martha Armstrong of the Massachusetts Society for the Prevention of Cruelty to Animals, Martin L. Stephens of the Humane Society of the United States, John Yam of Procter & Gamble, John Wilson of Johnson & Johnson, Oliver P. Flint and Wayne Carlson of Bristol-Myers, Robert Scala of Exxon, and Emil Pfitzer of Hoffman-LaRoche Pharmaceuticals. There was no hope of involving individuals and groups committed to outright abolition of animal use in science, or those scientists who believed that any discussion of laboratory animal welfare was a concession to animal rights claims. The former continued to push for all-or-nothing solutions, and the latter resisted suggestions that it was possible to improve the way animals were viewed and treated in laboratories or that such changes might be beneficial to science.

That latter group instead funded organizations whose goal was to counter the media-savvy animal rights groups' wooing of the public and legislators, often using the same tools of emotional manipulation that the animal rights groups had perfected over the previous decade. In place of posters of suffering animals, organizations like the Foundation for Biomedical Research (FBR) distributed photographs of beautiful healthy children, pointing out that their freedom from such formerly lethal childhood diseases as diphtheria and polio was the result of research on animals. Superimposed over a photograph of an animal rights demonstration, another poster's caption claimed, "Thanks to animal research, they'll be able to protest 30 years longer."

In 1985, the Association for Biomedical Research, a lobbying group supported by industry, and the National Society for Medical Research, which largely represented medical schools, had merged to form the National Association for Biomedical Research (NABR) and its sister organization, the Foundation for Biomedical Research. Headed by Frankie Trull, a former employee of animal breeder Charles River Laboratories, NABR represents more than three hundred research institutions and companies and holds yearly conferences in Washington, D.C., to keep its members informed on the latest animal rights challenges and threats. The group also lobbies Congress and works behind the scenes to counter animal protectionist legislation.

A number of state and local research defense groups were also formed in the wake of animal rights activity, including Connecticut United for Research Excellence (CURE), founded in 1990, which produces many publications and curricula for children, and similar groups in Pennsylvania, Oregon, California, North Carolina, New Jersey, and Wisconsin.

According to researchers at Tufts University Veterinary School, "The research advocacy groups together currently devote around $5 million a year to support the need for animal research. However, these funds do not include the activities of the professional scientific and medical societies, of the National Institutes of Health, or of the many corporations that are now actively engaged in the debate."

However, the relatively tasteful and measured approach of the FBR and others of these societies, supplemented by the congressional lobbying of the NABR, was viewed as too little too late by some individuals and organizations. Frankie Trull and Barbara Rich, director of the FBR, were savvy and articulate defenders of the need for animal research and worked hard to defend their organizations' constituencies from any more restrictive legislation. But animal rights groups appeared to be making inroads on American support of biomedical research, particularly among young people. Surveys conducted by the National Science Board revealed that in 1993 only 53 percent of Americans polled agreed with the statement "Scientists should be allowed to do research that causes pain and injury to animals like dogs and chimpanzees if it produces new information about health problems," compared with the 63 percent who agreed with the statement in 1985. The number who disagreed strongly with that statement rose from 30 percent to 42 percent over the same eight-year period. By 1996, only 50 percent of the respondents agreed with the statement, while 46 percent disagreed. Clearly, animal rights campaigns were having an effect, and just as clearly, a more aggressive approach by research defense groups was needed, something that could turn back the tide. It was time to go on the offensive.

LOS ANGELES, 1996

STAR-STUDDED celebrity awards shows are as common as cosmetic surgery in Hollywood. But PETA's 1996 gala, held at Paramount Studios to honor individuals in the entertainment business who were dedicated to advancing the cause of animal rights, was unique for reasons that its host and honorees would prefer to forget. As guests such as ex-Beatle Paul McCartney and his wife, Linda, longtime vegetarians who had been awarded PETA's Lifetime Achievement Award, Ellen DeGeneres, Mary Tyler Moore, Oliver Stone, and Woody Harrelson (all recipients of PETA Humanitarian Awards) entered the grounds, they were accosted by a mob of screaming protesters attacking everything the gala represented.

According to Susan E. Paris, director of Americans for Medical Progress (AMP), the research advocacy group sponsoring the demonstration, these celebrities and others like them had been duped. "PETA has sold Hollywood's best and brightest stars a bill of damaged goods," she said in a press release announcing the event. "These celebrities must understand that medical research is more than wearing a ribbon on your lapel. The only way we will be able to continue to develop new cures and treatment is through humane research on animals."

The same message is being disseminated by a number of state and local research defense groups throughout the United States. The thing that makes Americans for Medical Progress different is its tactics, torn straight from the pages of an activist's manual, and the fact that from its inception the organization has been heavily funded by the U.S. Surgical Corporation. Just as Mary Lou Sapone was hired to infiltrate the animal rights movement, a proresearch counterpart to the animal rights activists then infiltrating labs, so Susan Paris and her staff were funded to duke it out with the animal rights folks in the pages of the nation's newspapers and on the Internet. It's a public relations war now, and Paris and her director of public affairs, Jacqui Calnan, are top-notch propagandists—street fighters far more willing to get dirty than Barbara Rich and Frankie Trull, the elegantly dressed and coiffed doyennes of NABR and FBR. (In late 1998, Paris resigned and Calnan became director of the organization.)

AMP was founded in 1991 with a $900,000 donation from the U.S. Surgical Corporation to defend animal research more publicly and aggressively than did existing groups. At that time, U.S. Surgical was listed on the organization's letterhead as an affiliate of the group, although it provided all of AMP's funding. AMP's director, the attractive, aggressive Paris, described on the group's Web site as "the nation's foremost citizen advocate for biomedical research," is a former U.S. Surgical secretary who once worked as a personal assistant to Leon Hirsch. AMP's supporters in the research community were counting on her to conduct a high-profile war against animal rights and to convince both the public and celebrity supporters of PETA that they had been duped.

According to the 1996 numbers, support for research was continuing to erode. A pool taken by *Extra*, a syndicated tabloid news show, in conjunction with its coverage of the PETA Hollywood gala, revealed that 64 percent of viewers responding to a phone poll answered no when asked "Do you support the use of animals in medical research?" Although the respondents were far from a scientific sample, the poll's results concerned many who saw it as yet another indication of the success of animal rights groups in conveying their message to the public. "In light of this and other recent examples, members of the Alliance and those interested in maintaining the integrity of medical research should strengthen their commitment to inform people about the benefits of animal research to human and animal health," commented an editorial published by the Health, Safety and Research Alliance of New York State.

Americans for Medical Progress led the fight, although its tactics were not universally lauded by members of the research community. While decrying the targeting of individual researchers by animal rights groups, AMP publicly harassed Alec Baldwin and Kim Basinger, well-known supporters of animal rights. The day after the couple's child was born, AMP placed an ad in *Variety*, the Hollywood trade paper, addressed to Baldwin. The full-page ad depicted a baby on life support and said in bold type, "Dear Alec Baldwin and other supporters of animal rights: When you help PETA, this is who you hurt the most." In response, Baldwin called Paris "delusional" and added that "if you find it necessary to unfairly characterize my own position on animal rights, I will be forced to sue you until you are bleeding from the eyeballs. In future I suggest you communicate with me through my New York attorney."

At the Hollywood PETA gala, Paris and her group borrowed yet another tactic from the pages of the animal rights movement by holding their demonstration outside the event, held at Paramount Studios. AMP's newly developed skills in street theater were the result of its recent affiliation with another venerable group of demonstrators, the AIDS Coaliton to Unleash Power (ACT UP), the activist organization founded to demand greater responsiveness from the research community to the needs of AIDS patients. Like animal rights groups, ACT UP disrupted many a research conference in an attempt to call attention to its cause. However, as researchers and physicians became somewhat more responsive to the concerns of ACT UP's constituency and began to work more closely with patient advocacy groups, ACT UP chapters in many cities became more quiescent.

However, AMP's campaign against the "hypocrisy of Hollywood celebrities" who wear the red ribbon supporting AIDS research while donating time and money to PETA gave some members of ACT UP a new cause. "The greatest immediate threat to the lives of people with HIV and AIDS is the antiresearch agenda of the animal rights groups gathering in Washing-

ton, D.C., this week," said a June 1996 AMP press release, referring to the annual Animal Rights Conference held in Washington
each spring. A letter-writing campaign to U.S. newspapers, and
statements by AMP spokespeople and board members, made the
same point.

While AMP accuses PETA of using and duping Hollywood
celebrities to advance the animal rights cause, AMP's use of ACT
UP and AIDS patients has sometimes seemed equally cynical.
Jeff Getty, the thirty-eight-year-old man who received a baboon
bone marrow transplant in 1996 in an effort to boost his immune
system, is one example. This procedure took place despite reservations expressed by many virologists that xenotransplants, in
which tissues and organs are transferred between species, pose
the risk of spreading lethal animal viruses to the human population, particularly when primate organs and tissues are used.
Within weeks of the surgery, AMP had adopted Getty as its
favorite patient spokesperson and used him to attack animal
rights in the pages of *The Wall Street Journal* and other media. "It
is not enough for us to wear red ribbons and hope that a cure is
found," said Joseph Murray, a Nobel laureate and board member
of AMP. "We must actively support those scientists, doctors, and
brave volunteers, such as Jeff Getty, who are on the front lines of
research."

Meanwhile, ACT UP and other advocates of people with
AIDS, goaded by AMP, were infuriated by callous comments by
PETA spokespeople about the disease. In a 1994 article in *New
York* magazine, PETA's director of special projects, Dan Mathews, was asked about PETA's opposition to animal research in
the face of the lethal epidemic. "No AIDS breakthroughs have
come out of animal research," Mathews told the interviewer.
When a spokesperson for the National Institutes of Health contacted by the author of the article called this assertion nonsense,
Mathews remained (in the words of the interviewer) "unrepen-
tant," commenting that "we live in a lazy, sick society" and that
"people bring disease on themselves." As indicated in this exam-

ple, PETA and many other animal rights and antivivisection groups continued throughout the nineties to deny any benefit from animal experimentation and to assert that, in any case, possible benefits do not outweigh animal suffering.

In 1996, however, AMP targeted not only abolitionists like PETA but more moderate groups as well in its efforts to turn the tide. In November, it launched a broadside attack on the Humane Society of the United States, one of the nation's wealthiest and most powerful humane groups.

"Regardless of the name they use—PETA, the Humane Society of the United States, or the PETA-fabricated front group Physicians' Committee for Responsible Medicine (PCRM)—animal rights zealots have a single objective: stopping the use of animals in biomedical research. If they succeed, AIDS research will be crippled and hundreds of thousands of people will suffer needlessly and die," Paris said in a June 18, 1996, press release. In November of that year, AMP launched a mass mailing of material stating that "for the past two decades, concern over the threat to the biomedical research community by the animal rights movement was centered on groups such as People for the Ethical Treatment of Animals (PETA), the Animal Liberation Front, and the thirty or so other organizations that openly advocate radical opposition to scientists working with animals. During this time, the Humane Society of the United States (HSUS) has worked methodically to gain a great deal of access to the public, the media and the research community because of its supposedly moderate public stance. The fact is that the Humane Society of the United States, including its personnel and programs, reveals a portrait of a hardcore animal rights group positioned by virtue of its wealth and carefully crafted national and international affiliations to do more damage to biomedical research than those organizations that openly espouse the radical animal rights ideology and agenda."

AMP's information sheet titled "HSUS-PETA connections" listed only one HSUS senior staffer, Rick Swain, a twenty-five-year police veteran. Swain had joined PETA's staff after his retirement, serving as managing director, before leaving to become vice president of investigations for HSUS. The other individuals on the list were tainted by personal associations with PETA staffers rather than professional ties—for example, Wayne Pacelle, vice president for media and government affairs at HSUS, was described as a "friend" of PETA cofounder Alex Pacheco. A number of low-level HSUS staffers were listed and described as former PETA employees. The AMP summary did not discuss their reasons for leaving PETA, but it is possible that rather than "infiltrating" HSUS, these individuals were attracted by its higher salaries, more moderate stance, and more congenial working conditions.

AMP and other research defense groups maintain that there is an unbridgeable chasm between animal welfare and animal rights—an assertion identical to that made by proponents of animal rights orthodoxy. But a look at the membership and perspectives of groups like the Humane Society of the United States and the American Society for the Prevention of Cruelty to Animals, the nation's two largest animal protection groups, both of which support a wide range of programs and neither of which is explicitly antivivisectionist, illustrates the problematic nature of that assertion. A broad range of perspectives coexists within the movement, and the contemporary continuum of animal protection stretches from the meat-eating conservationists who donate to groups like the HSUS, the ASPCA, and the Sierra Club to the raiders of the ALF, with every shade of ideology and practice in between.

Nonetheless, whether one categorizes them as "welfarists" or "rightists" it is clear that many Americans, and even more citizens of northern European countries such as the Netherlands, Britain, and Germany, are questioning long-held beliefs about the human-animal relationship. Opinion surveys carried out over the past fifteen years have charted substantial shifts in public attitudes toward animal use, although surveys also reveal a divided response to the animal rights movement per se. Psychologist Harold A. Herzog, Jr., who has studied both the movement itself and public response to animal activism, reports that public attitudes toward the animal rights movement are mixed. He cites a 1994 opinion poll that showed "that most respondents had either a very favorable (23%) or mostly favorable (42%) view of the animal rights movement." Nonetheless, Herzog notes that "only 7% of a 1990 survey said that they agreed with both the agenda of the animal rights movement and its strategies. Eighty-nine percent of the respondents felt that activists were well meaning, but either disagreed with the movement's positions on issues or on strategies for accomplishing specific goals."

Public distaste for the movement's strategies and discomfort with the implications of a radical animal rights agenda have grown as the absolutist wing of the movement has garnered more media attention in recent years. Animal rights "death lists" in Britain, which target individual scientists for their use of animals, and sustained campaigns of harassment against scientists by some groups in the United States have tainted the entire movement. As more pragmatic activists work quietly within the system for change, the aggressive acts of extremists have assumed greater prominence. Although the ALF has always explicitly worked to intimidate scientists, the latest round of harassment by campus groups and national organizations has brought such intimidation out of the laboratory closet and into the community. These campaigns inevitably create sympathy for besieged researchers and their families and cause many members of the public, who may initially have been sympathetic to certain elements of the animal rights agenda, to question the movement's commitment to the very principles of compassion and ethical action it purports to champion.

EIGHT

SAINTS AND SINNERS

Many in the movement endorse "diversity" and urge that anyone who "cares" for animals, or who has "compassion" for animals, is really "walking along the same road" with those whose long-term goal is the abolition of animal exploitation. This desire to embrace "diversity" in the movement leads to positions that are difficult to understand and that make it difficult to formulate criteria for distinguishing who is the "exploiter" and who is not.

—Gary L. Francione,
Rain Without Thunder:
The Ideology of the Animal Rights
Movement, 1996

MINNESOTA, 1996

HALLOWEEN. The streets in Marilyn Carroll's suburban neighborhood are full of children in costume, ringing doorbells and shouting, "Trick or treat!" Late that evening, a group of adults knocks on Carroll's door. Her seventeen-year-old daughter answers, candy in hand, and is confronted by a group of protesters. One of the signs carried by the group is shaped like a tombstone and says YOU WILL NEVER R.I.P., MARILYN THE VIVISECTOR. The group shouts at the girl, "Your mother kills animals," and the masked individuals on the porch try to press flyers into her hand. Frightened, she calls for her mother.

When Carroll herself approaches the door, the protesters begin shouting, and one of them pulls out a camera and tries to take her picture. "As I turned away and walked back into the interior of my home, the protesters chanted, 'Marilyn, come here, we want to talk to you; stop killing animals,'" Carroll told the Foundation for Biomedical Research in an interview published in the March 1997 issue of *FBR Facts*. "My husband then came to the door to tell the protesters to leave. They asked him if he enjoyed torturing animals. He responded by asking one of the male protesters why he was covering his face with a black cloth. The protester said because he was a member of the ALF. Then another male in the crowd said something about burning the house down. My husband said, 'Excuse me, what did you say?' The protester stated that the pumpkins were smoking and if we weren't careful, the pumpkins might burn the house down. We considered this a threat and felt very intimidated and harassed by the actions of the protesters."

The Halloween incident was the outcome of a course of events set in motion more than a year earlier, when Freeman Wicklund, a twenty-three-year-old activist and member of the Student Organization for Animal Rights (SOAR), occupied the office of the president of the University of Minnesota to protest Carroll's use of rhesus macaques and rodents in addiction studies funded by the National Institutes of Health and the National Institute on Drug Abuse. Arrested and sentenced to a year's probation for disorderly conduct and trespassing, Wicklund asked to serve a ninety-day jail sentence instead. The day of his sentencing, March 5, 1997, Wicklund launched a hunger strike, which he told reporters he intended to continue throughout his sentence. Two weeks later the judge who had agreed to the ninety-day sentence in lieu of a year's probation "said she had misread the sentencing laws and decided Wicklund did not have the option to reject probation for jail time. Wicklund was released and stopped his hunger strike," according to the FBR.

Undeterred, Wicklund and his colleagues at SOAR stepped up their campaign against Carroll in September 1996. Flyers distributed at the UM campus on September 26, the first day of the fall semester, identified Carroll as "Vivisector of the Month" and charged that "Marilyn Carroll's project at the U. of M., 'A Primate Model of Drug Abuse,' has been torturing animals for over eighteen years under the protective blanket of science." The fly-ers encouraged those opposed to this research to "write, call, fax or email Marilyn Carroll and tell her what she's doing is useless and cruel" and provided her e-mail address, fax number, and both office and home telephone numbers and addresses.

That same day, "five activists came to my laboratory in Diehl Hall. They carried a video camera and attempted to record activ-ities in my laboratory," Carroll told *FBR Facts*: "We were forced to lock the doors of the laboratory, computer room, and two animal rooms and shut down our research activities until the protesters left. They dispersed after being informed that police had been called." Later that afternoon, Carroll received a phone call from a young woman who "said that she did not like what I was doing to my animals and stated, 'How would you like it if you were killed like you are killing your research animals?' "

On October 8, Carroll and her family awoke to find a group of protesters "standing at the foot of our driveway carrying large signs (about two feet by three feet) and a painted sheet-sign protesting my research activities. They rang our doorbell, but we did not answer the door. We called the police because we needed to leave the house to take our youngest child to school and go to work ourselves, but we did not wish to drive through the protest-ers blocking our driveway." The police arrived and the crowd dis-persed, after identifying themselves as members of SOAR.

Carroll and her family returned home at dinnertime that same day, October 8, to find another group of protesters in the drive-way "carrying large signs and a painted sheet . . . There was a large group of about ten neighborhood children in our yard rang-ing in age from about 3 to 17. My neighbor across the street had called the police while we were away to complain that the pro-testers were harassing the children, making comments such as 'your neighbor kills animals.' The protester had distributed fliers to the children which identified their organization as SOAR." According to *FBR Facts*, Carroll received a number of threaten-ing letters and telephone calls during this period. Her neighbors were also harassed and were "enraged because activists handed out literature containing disturbing photos of animals to their young children." In November, Carroll secured a restraining order against several of the protesters and against the organizations they claimed to represent, the ALF and SOAR.

Freeman Wicklund, one of the protesters identified in the restraining order, has been identified by the National Association for Biomedical Research as "the most visible of a new generation of animal activists who, dissatisfied with 'traditional' animal rights strategies, promote an agenda of radical activism." In 1995, Wicklund had organized an ALF Appreciation Day, and at the annual World Congress for Animals in Washington, D.C., in June 1996, he called for support within the movement for ALF activities, referring to members of the organization as "selfless soldiers," "compassionate commandos," and "juggernauts of justice," according to a crisis management manual distributed by NABR. Attorney Jo Shoesmith, who attended the congress, said that Wicklund's presentation was one of the most popular at the meeting.

BY THE MID-NINETIES, the animal rights movement's approach to biomedical research had diversified greatly. Some individuals and groups concluded that the protests of the previous two decades had effectively set in motion an institutional reform movement within science that was proceeding on its own momentum, and turned their attention to other issues. Henry Spira, for one, had largely turned away from laboratories and begun to focus on factory farming as early as 1986, noting that the number of animals suffering as a result of modern agricultural practices far exceeded the number of animals suffering in laboratories and that "as long as six billion animals are being consumed as food, I don't think you're going to have a point where there's not even one animal being used for product testing."

As Peter Singer noted in his biography of Spira, explaining the latter's shift to protesting agricultural uses of animals: "While the American animal movement reached new levels of public awareness in the first half of the 1980's, it was narrowly focused on animals used in research, with some concern for stray dogs and cats and for wildlife. Farm animals were almost entirely neglected. Yet for each one of the 20 to 60 million animals then used in research in the United States each year, at least 200 farm animals were killed . . . Moreover, whereas the number of animals used in research seems to be falling, the number of farm animals is rising rapidly."

Other activists began to work for increased legal protection for laboratory animals and a more substantive commitment to the three R's from the research community. The Animal Legal Defense Fund, the Humane Society of the United States, and two individuals brought suit in 1990 to petition the USDA to include rats, mice, and birds as protected animals under the Animal Welfare Act. The ALDF won the case on January 8, 1992, when a U.S. district court judge, Charles R. Richey, ruled that the USDA's exclusion of rats, mice, and birds was arbitrary and capricious. (This decision was overturned on appeal, however.) The ALDF also won a 1996 case charging that the USDA had failed to implement the 1985 requirements of the Animal Welfare Act calling for federal guidelines to ensure the psychological well-being of primates and the proper exercising of dogs. (Again, this ruling was overturned by an appeals court that decided the ALDF did not have legal standing to sue on behalf of laboratory animals.)

Despite these setbacks, a number of animal protection groups, including those explicitly devoted to the abolition of animal experimentation, have shown a willingness to communicate with the research community to improve conditions for laboratory animals. For example, key staff members of the American Anti-Vivisection Society (Tina Nelson), In Defense of Animals (Suzanne Roy), and the Animal Legal Defense Fund (Valerie Stanley) attended either the 1996 or the 1997 annual meeting of Public Responsibility in Medicine and Research (PRIMR), a Boston-based educational forum founded in 1973 to help research institutions understand and implement the increasing number of state and national laws regulating biomedical research.

Although representatives of more moderate groups, such as Martin L. Stephens of HSUS, and Cathy Liss of the Animal Welfare Institute, had been attending these meetings for years, the invitation to Suzanne Roy and Henry Spira to speak at the March 1997 meeting was controversial, and some researchers threatened to boycott the meeting. But PRIMR's director, attorney Joan Rachlin, who says that PRIMR "values highly our interdisciplinary character," refused to back down and the meeting proceeded as planned.

As the three R's approach has spread throughout industry, regulatory agencies, and the research community, giving scientists and animal protectionists a common language and a vehicle for identifying common goals, this type of boundary crossing has become more widespread. By fall 1998, even Mary Beth Sweetland, director of investigations for PETA, the perceived archnemesis of the research community, had attended an FDA meeting on alternative testing, sitting around the table with FDA officials and representatives of the Colgate-Palmolive Company to discuss possible areas for cooperation. "I was a bit concerned about inviting her to the meeting because of the group's reputation, but she was very cordial and even asked "What can we do to help?" said Neil L. Wilcox, then a senior science policy officer in the Office of the Commissioner, who has been instrumental in encouraging the agency to develop alternatives to traditional whole animal models for testing.

The increasing willingness of some activists and organizations to work within the system for change has been attacked by others who sense that the aggressive rhetoric and all-or-nothing demands of the early years of the movement are being replaced by pragmatism and flexibility even among groups that claim to be radical supporters of animal rights. The most fervid proponent of the need to enforce a rigid animal rights orthodoxy is Gary L. Francione, professor of law at the Rutgers University Law School and codirector (with his wife, Anna Charlton) of the Rutgers Animal Rights Law Center (now defunct). In *Rain Without Thunder: The Ideology of the Animal Rights Movement*, Francione attacks a number of high-profile activists and theorists, including Peter Singer, Henry Spira, Ingrid Newkirk, and Kim Stallwood, editor of *Animals Agenda* magazine, for betraying the goals of the movement. Calling the more moderate approach developed throughout the nineties the "new welfarism," Francione argues that true "advocates of animal rights are not interested in *regulating* animal exploitation, but in *abolishing* it" and charges that "once we abandon animal rights idealism in favor of a standard that requires only that we 'care' or 'feel compassion' toward other animals, it becomes impossible any longer to differentiate animal rights theory from welfarist notions that are accepted by virtually everyone—*including animal exploiters.*"

Francione posits that an ideological chasm separates animal rights and animal welfare and that "the rejection of rights theory by supposed rights activists is becoming increasingly apparent." He compares the style and rhetoric of the 1990 March for Animal Rights with those of a similar event held in June 1996 and notes that "the tone of the 1996 march is clearly more moderate than that of the 1990 march, and it reflects the deliberate and explicit rejection of animal rights by many animal advocacy groups." Charging that once-radical groups like PETA, and once-radical activists like Spira, had allowed themselves to be co-opted by welfarism, Francione calls for a return to a pure animal rights theory and praxis, one characterized by a rejection of "instrumentalism" in all its forms. Francione rejects Peter Singer's utilitarian philosophy as a basis for animal rights activism—going so far as to challenge Singer's status as "father" of the modern animal rights movement—and locates the roots of the new welfarism in Singer's work. Instead, he argues in favor of Tom Regan's deontological philosophy of animal rights, which he believes provides a more solid grounding for the rejection of instrumentalism.

"The foundation of animal rights theory is the elimination of the property status of animals," Francione says in *Rain Without*

Thunder, arguing that animal rights theory "presents an argument for the abolition of institutionalized animal exploitation." Francione, following Regan, writes that "it is wrong to treat animals in a completely instrumental way, just as it is wrong to treat humans in a completely instrumental way. And it is wrong because, at the least, animals that are subjects-of-a-life have inherent value, and they have it because all subjects-of-a-life are relevantly similar. There is simply no nonspeciesist way of differentiating human subjects-of-a-life from nonhuman ones, which have inherent value for precisely the same reason that the humans do: because their life matters to them apart from whether it matters to anyone else."

Francione concludes that when conflict arises between human interests and the rights of an animal as a subject-of-a-life, the rights of the animal should prevail—a complete reversal of traditional views and a transformation that can be accomplished only if animals are no longer viewed as human property under law. As long as animals are legally treated as human property, he maintains, their rights will always be perceived as secondary to human needs and interests. "The trouble with property status for animals is that any interest that is recognized is subject to sacrifice and that, barring unusual and exceptional circumstances, these interests are always expendable as long as the requisite benefit is found. So, as a conceptual matter, it seems as though any incremental eradication of property status must involve interests that are not expendable *even if there is significant human benefit to be derived from ignoring the animal's interest.*"

This pure ideology of animal rights as defined by Francione would prohibit all forms of animal research and testing on animals deemed "subjects-of-a-life." Using the terms of Francione's argument, the strength of the legal and moral prohibition against killing or using another animal for one's own ends would be equal to the strength of the moral prohibition against killing or using another human. Although this Golden Rule has been rather spottily implemented in human-human interactions, Francione rejects the notion that extending such protection to animals and expecting humans to honor it constitutes a utopian vision. Perhaps that is one reason nearly everyone in the animal protection movement except Francione himself and the late Helen Jones, founder of the International Society for Animal Rights, falls short, in his estimation, in their devotion to the movement's ideals.

Francione (dubbed "the middle-aged *enfant terrible* of the animal rights movement" by journalist Merritt Clifton, editor of the animal rights newspaper *Animal People*) appears to be engaged in a familiar form of radical theater. Like all revolutionary movements, animal rights seems to have reached the cannibalistic stage in which pitched battles for ideological purity cloak a struggle for power. Just as the radical Jacobins purged the more moderate Girondists after the French Revolution, and the Bolsheviks made short work of the Trotskyites in postrevolutionary Russia, so Francione and young activists like Freeman Wicklund seem determined to outgun the old guard. Of course, such power struggles can only take place after the first critical battles against the forces of reaction have already been won—a point that Francione refuses to concede. Dismissing the increased regulation of biomedical research and the moderate but measurable success of the three R's approach as so much welfarist obscurantism, Francione titles his final chapter "Marching Backwards." "Perhaps it is time to recognize that welfarist reforms lead to more animal exploitation, not less," he says. But recent events have shown that some things have changed quite a lot since 1981, when Edward Taub's laboratory was raided by police in Silver Spring, Maryland.

OMAHA, NEBRASKA, AUGUST 1996

ON AUGUST 14, Ed Walsh and JoAnn McGee, husband-and-wife researchers at the Boys Town National Research Hospital, found themselves in the news. PETA had held a press conference earlier that day alleging that a whistle-blower's complaint, followed by a seven-month undercover investigation at Boys Town by PETA, had revealed serious abuses of cats in Walsh and McGee's laboratories. "Prompted by a call from an employee who said he could no longer stand to see kittens and cats suffering in experiments conducted by federally funded researchers at Boys Town, PETA sent two investigators who documented what PETA believes to be serious violations of federal law in Boys Town's laboratories," said the PETA press release. The group filed a forty-four-page complaint with the U.S. Department of Agriculture and the National Institutes of Health alleging numerous violations of the Animal Welfare Act in Walsh and McGee's laboratories, including failure to provide adequate veterinary care to the animals, failure to euthanize suffering animals, and failure to properly train and supervise laboratory staff.

Walsh and McGee are auditory researchers who have studied the causes and possible prevention of congenital deafness since their graduate student days. "We knew that the cat inner ear underwent extensive developmental change during the first few weeks of life, changes that human fetuses undergo during the second and third trimesters of pregnancy," Walsh told an FBR interviewer in March 1997, describing his research program. "One key event that occurs during this process is the innervation of the inner ear by a brain pathway known as the olivocochlear bundle. We found it interesting and potentially important that the acquisition of inner-ear properties essential to normal hearing occurred in connection with the growth and organization of olivocochlear bundle projections. We wanted to figure out whether or not this brain projection induced the development of inner-ear elements that lead to the acquisition of normal hearing."

Noting that "domestic cats have been used to study auditory anatomy and physiology for many decades" and that "we know a tremendous amount about their anatomy and physiology and precious little about the development of auditory function in most other mammals," Walsh and McGee settled on cats as a research model. All the cats in their laboratory were purpose-bred—raised for research by animal breeders. Asked by *FBR News* to describe the experimental procedures denounced by PETA, Walsh complied: "The experiment is conducted in two steps. In step one, the olivocochlear bundle is severed in newborn cats before projections from the bundle become organized within the ear. This delicate operation is a major neurosurgical procedure and is conducted in an operating room using standard sterile procedures in deeply anesthetized subjects. The second step is to assess inner-ear and peripheral nerve function in experimental animals that were reared in the absence of the bundle's influence. We do this by recording the electrical activity produced by individual auditory nerve fibers in response to sound and comparing their response properties with those of normal and control subjects. The animals are deeply anesthetized throughout each and every session, ensuring that they experience no discomfort even though the procedure is invasive."

Walsh and McGee discovered that the ears of the cats whose olivocochlear nerves had been severed early in infancy "were profoundly abnormal" compared with those of control animals. "This finding suggests that the olivocochlear bundle either plays a primary role in the development of the inner ear, or its absence during development renders the inner ear hypersusceptible to traumatic environmental agents." Walsh believes that this research could "contribute to the understanding of brain malfunction during development in general, not just in relationship to hearing. All aspects of neuroscience, in my view, are served by studying audition, since many, if not most, of the principles operating in one brain area hold true in others."

The Walsh-McGee experiments were immediately suspended by Boys Town after PETA filed its complaint, and the institution initiated both an internal and external review of the Walsh-McGee research program. Six days after receiving PETA's complaint, the USDA sent inspectors to the laboratory for an unannounced examination. An External Review Committee composed of nationally recognized experts in auditory research spent two days at the laboratory in early September, reviewing the facilities and records. Laboratory staff were interviewed by USDA inspectors and members of the review committee. Both the USDA inspectors and the External Review Committee filed reports stating that they had found no evidence of animal abuse or maltreatment, but both recommended better record keeping and documentation of experiments. On September 30, three weeks after the PETA complaint was filed, USDA inspectors appeared at the lab for a second unannounced examination. Once again, they found no evidence of violations of the Animal Welfare Act and further noted that the earlier recommendations for improved documentation had been implemented.

In the meantime, local papers covering the story discovered that the "whistle-blower" identified by PETA in its original complaint was actually two people, "both of whom were PETA employees posing as Boys Town employees," Walsh said in a 1998 interview with the author. "Matt someone (I can't recall his last name) and Michelle Rokke, who had a documentable history of infiltrating other institutions, as well as our place. There is no evidence that anyone from my lab lodged a complaint with PETA or anyone else. To the contrary, all lab employees and students continue to support the lab, as they always have."

"That Boys Town was actually targeted by PETA, as opposed to whistle-blowing, was revealed in an interview with Mary Beth Sweetland that appeared in a local newspaper," he added. "Sweetland disclosed in an interview that PETA had used two sources to learn about Boys Town's research program involving cats. In addition to the Freedom of Information Act route, PETA discovered that cats were used in our research program through purchase order records that were stolen from the files of an animal vendor that they had infiltrated and with whom we had placed an order."

Like Marilyn Carroll, the Minnesota addictions researcher, Walsh and McGee were subjected to extensive harassment at both their workplace and home. By publishing the home telephone numbers and addresses of researchers, organizations like

SOAR and PETA virtually ensure that the researchers will be subjected to threatening calls, letters, and visits by individuals who object to their work as depicted in the graphic and disturbing flyers and videotapes purporting to describe the research. "I can tell you that aside from a serious illness that threatened our son's life when he was a toddler, I can't think of another incident as harmful, in both professional and emotional terms, as the intrusion of PETA into our lives," Welsh said in 1998. "Much to PETA's delight, I'm sure, simple acts like starting our car have taken on an air of danger, and hardly a day goes by that we aren't anxious about tomorrow."

Walsh and McGee's home has been picketed a number of times, and Walsh's elderly mother, alone with her grandson in their home, has been threatened. Even after their research program was cleared by NIH, USDA, and the External Review Committee put together by Boys Town, Walsh and McGee continued to serve as exemplars of cruel and irresponsible researchers in PETA's marketing materials. For example, at the June 1997 World Congress on Animals in Washington, D.C., over a year after these review committees had testified that the Walsh-McGee laboratories were in compliance with all federal regulations, the PETA booth at the conference was stocked with flyers and leaflets denouncing the research. As a result of the continuing harassment, Walsh and McGee decided to terminate the cat experiments and switch to rodent studies.

"As a consequence of their [PETA's] recklessness and their disregard for honesty and fair play, our professional and personal lives have been profoundly altered, even though their allegations were discredited by no less than three outside sources," Walsh said in September 1998. "Regardless, much to PETA's delight, I'm sure, a productive twenty-year biomedical research program was decimated as a consequence." When asked about the emotional costs of PETA's campaign, Walsh felt that "the distress produced by it was directly proportional to the extent that it was false, and it was fundamentally false. The evidence for that is ample, if one is inclined to look for it. Ingrid Newkirk or Mary

Beth Sweetland or whoever is assigned to the case today would undoubtedly argue that experimenting on and sacrificing our research animals is a strange way to demonstrate our concern for their welfare. The answer to such an argument is apparent and simple. Much as we care for our animals, and the extent of that care is far-reaching and delivered genuinely, we care about the suffering of mankind more. While we always minimize, if not eliminate, suffering, even discomfort, in our subjects, we are fundamentally dedicated to advancing medicine and civilization itself. Always anchored to a solid ethical bedrock, the use of animals is an essential element in that cause."

Walsh vehemently denies that PETA's campaign has caused him to "reevaluate my beliefs or change my practices for one simple reason—reevaluating the moral basis of using animals in my research is part of the process of conducting research." On the other hand, he says, "I was and continue to be consumed by the need to understand malevolence in society after experiencing its wrath. I can hardly believe that this group could so casually misrepresent me and my research. As a result, I was driven to reconsider the relationship between the first amendment and the right of privacy, to consider the ethical basis of special-interest dynamics. In the end, I rediscovered what I already knew. The indignity and persecution of an individual, and in this case his family, is a small price to pay for liberty, for the freedom that resonates at the center of an open society."

Nonetheless, he remains baffled by the response of institutional authorities to the PETA charges. "I continue to be troubled by the position taken by responsible government agencies. A guilty-until-you-prove-yourself-innocent stance is counterintuitive to the free society that we seek. I hope to live to see the day that such regressive thinking is turned around."

THE EXPERIENCE of Ed Walsh and JoAnn McGee is emblematic of the changes that have taken place in the regulation of research in U.S. laboratories—and in the institutional response to charges of animal abuse. A regulatory and investigative apparatus is now in place that far exceeds anything that might have been imagined twenty years ago. Recall that it took eighteen months of public protest, with extensive media coverage and congressional pressure, to end arguably trivial experiments on cats at the Museum of Natural History. In 1996, a complaint filed by a single animal rights group resulted in the immediate cessation of a serious study of a medical problem at a well-known and respected institute and a rapid response and thorough investigation by regulatory authorities.

Even before beginning their experiments, Ed Walsh and JoAnn McGee (like all other federally funded investigators in the country) were required by federal law to seek the approval of their institution's animal care and use committee (IACUC) for the projected experiments before seeking funding. The 1985 amendments to the Animal Welfare Act, passed in the wake of the University of Pennsylvania Head Injury Clinic controversy, mandated the establishment of IACUCs in all research institutions. IACUCs must comprise at least five members, including a doctor of veterinary medicine, an experienced investigator, a non-scientific member (usually an ethicist, lawyer, or member of the clergy), and an "unaffiliated" member who has no ties to the institution. Many IACUCs comprise more than the required five members.

IACUCs are responsible for reviewing all scientific protocols requiring animals and are expected to scrutinize the proposed experiments carefully, judging not the scientific merit of the experiment but the proposed uses of animals. Veterinarians, both in-house and those in private practice, who often serve as unaffiliated members, are key components of the committee. "While serving as a member of the committee, I tried to insure that the number of animals used was not excessive, and that the animals were protected from as much pain and stress as possible," said John Fioramonti, a veterinarian in private practice who served as an unaffiliated member of an IACUC at a large research institute in Baltimore. "As a veterinarian, I knew what drugs were available for anesthesia and analgesia and what the correct dosages were. There were two other veterinarians on the committee as well, so that any time a member of the committee had a question about procedures, there was a vet available to answer the questions."

One of the more controversial provisions of the 1985 legislation was the IACUCs' mandate to "determine that the investigator has considered alternatives to procedures that may cause more than slight pain or distress to the animals, and has provided a written narrative of the methods and sources used to determine that alternatives were not available." In 1986, an information service was set up at the National Agricultural Library in Beltsville, Maryland, to assist investigators and IACUCs in carrying out their responsibilities. "While the regulations seem fairly straightforward, it has been our observation that many people are unsure exactly what an alternative is and are confused as to what information is required to show compliance," Tim Allen and D'Anna Jensen, two of Animal Welfare Information Center's information specialists, wrote in a 1996 article titled "IACUCs and AWIC: The Search for Alternatives." "There are many opportunities to incorporate alternatives into an experimental procedure; however, many IACUCs and scientists mistakenly assume that only nonanimal methods satisfy the definition of an alternative."

This is not the case, as Allen and Jensen note in their article. "The U.S. Department of Agriculture views alternatives with an eye to the 3Rs so eloquently described by W. M. S. Russell and R. L. Burch in *The Principles of Humane Experimental Technique*—reduction of animal numbers, refined procedures to minimize or avoid pain, and replacement of animals with non-animal models," they said. To help investigators and IACUCs understand the new requirements, AWIC developed a two-day workshop called Meeting the Information Requirements of the Animal Welfare Act, which provides an "overview of the Animal Welfare Act, federally mandated IACUC functions, criteria for granting IACUC approval for animal research, and the required contents of institutional training programs" as well as an introduction to the concept of alternatives and information on the use of multiple databases and search strategies for finding alternatives. AWIC staff conduct the workshops at research institutions and at the National Agricultural Library in Beltsville. Nearly three thousand researchers and IACUC members have attended the workshops since 1991, some of them from as far away as Britain, Germany, Canada, and Puerto Rico, and "the most common sentiment [expressed by participants] is that the class should be required for all members of IACUCs, as it addresses many of the problems common to successful IACUC functioning."

The information specialists at AWIC also began publishing a newsletter in 1990, distributing CD-ROMs and bibliographies, and attending scientific meetings to interact with the research community and introduce them to the service. "Comments at meetings or workshops reveal that many scientists and IACUC members view the alternatives search as unnecessary government intrusion into the research process, and not as a resource that might enhance or improve their research," Allen and Jensen note. Investigator resistance to IACUC review has also been described by Dartmouth toxicologist and IACUC chair Bill D. Roebuck in an article published in the Center for Alternatives to Animal Testing newsletter in 1996 and in an interview in 1997 in *The Scientist*, a newspaper for researchers. Attributing much of the resistance to time constraints and the tendency to view IACUC review as just another bureaucratic hurdle the investigator needs to jump in order to do research, Roebuck nonetheless admitted that "most of the negative reaction to IACUC review results from the investigator's perception of the review as an insult. 'A committee is questioning my ideas and methods.' Perhaps the idea of review by a committee, some or most of whom are unknown to the investigator, is more upsetting than if the reviewer were a knowledgeable and respected associate suggesting that an experimenter do an experiment differently. Adding to the insult is the almost certain knowledge that the investigator knows far more about the grant or project than anyone on the IACUC."

Roebuck said that the way around this impasse is to "get the investigator to view the IACUC review as an opportunity—a research opportunity. IACUC review offers the investigator the opportunity to review his or her research plans, the opportunity to confront some difficult scientific choices, and the opportunity to evaluate some new or alternative choices." Improved relationships between investigators and IACUC members are the key to the process, he said. "Intellectual engagement is the key to developing the relationship. Half-hearted efforts or simplistic approaches will not work. Most important, "A relationship of trust or respect between veterinarians and investigators is important. . . . I believe that one of the best ways to build such relationships of trust is for the veterinary staff to serve as keepers of important knowledge of techniques and procedures involving animals. This knowledge can only be acquired by working actively with investigators, manually and intellectually."

In the same article in *The Scientist*, "Veterinarians in Research Labs Address Conflicting Agendas," veterinarian Phillip C. Tillman of the University of California at Davis said that "researchers don't see why [the IACUC] is obstructing them and making them fill out all these silly forms . . . they're scientists and they have an inspiration. It came to them in their sleep last night. They want to start on it today. And they just can't do that anymore. . . . It takes at least a month between the time that you get the idea and when you can get a committee to approve it." That time lag may be substantially increased if the IACUC objects to any element of the protocol and sends it back to the researcher for modifications. The authority of the IACUC to recommend and enforce modifications is not always welcomed, even when the changes are relatively minor. However, IACUC members have a responsibility not only to the animals but to the institution, and a failure to scrutinize protocols to ensure that they are in compliance with USDA and Public Health Service regulations could have disastrous consequences.

For that reason, many supporters of animal rights believe that IACUCs rubberstamp protocols and are more interested in facilitating the work of investigators than in protecting animals. "The ALF . . . thinks you're a pimp for the research community," Tillman commented in *The Scientist*. Admitting that he had once found a dead fish on his doorstep after he appeared on television, he went on, "I've often said that if you get approximately equal stacks of hate mail from investigators and from animal rights groups, you're probably doing a pretty good job."

Some of the resistance to the continuing efforts to improve the status of laboratory animals in the United States might be mitigated if American researchers were better informed of the scientific rationale for the three R's approach. British researcher Michael Chance discovered decades ago that environmental conditions that affect the behavior of animals also affect animal physiology, reporting that the ovary weight of immature female rats exposed to the sex hormone gonadotropin varied greatly according to the conditions in which the animals were housed. "The variation in ovary weight (the test response) was *greater* if the cages were small and cramped, if there was frequent disturbance by changing cages and cage-mates, and if the rats were caged either singly or in groups larger than two (with the floor area per rat roughly constant)."

This finding led Chance to review studies he had conducted during the 1940s on drug toxicity and potency. "In both cases, Chance found notable effects of environmental conditions on the variance of animal's test responses. . . . In 1962, Chance's colleague John Mackintosh reported a study of the responses of male mice to a barbiturate anesthetic, and he showed that the variance of the response was greater if the animals were caged either singly or in groups of eight than if they were caged in pairs." Further research led Chance and colleagues to conclude that species-inappropriate housing and handling, excessive noise, and other environmental variables "have psychosomatic physiological effects which must disturb any experimental results, and will thus make tests unreliable when repeated or performed in different laboratories." The researchers reported a number of provocative findings, including a higher incidence of cancer in mice in overcrowded cages, variable blood cell levels in mice exposed to disturbing sounds, and susceptibility of rats to tuberculosis increased by overcrowding.

In their 1997 paper, Chance and Russell concluded that "if we use ethological sophistication to provide laboratory animals with the very best physical and social environmental conditions for their well-being, we need to use fewer of them in research experiments or routine tests, and our results will be accurate and reliable." Over the past twenty years, more scientists have begun to appreciate the impact of laboratory conditions on animal well-being, but the mind-set that sees animals as laboratory tools and not sentient beings, as sensitive as humans to uncomfortable or unpleasant surroundings, fails to see the value in a more empathetic (and scientific) approach to animal husbandry. This is particularly true when it becomes clear that paying more attention to refinement issues will drive up the cost of doing animal research—a trend that has been building steadily and shows no signs of abating.

In a 1997 policy article in *Science,* a number of veterinarians chairing departments of comparative medicine at major research universities noted the rising cost of animal research and recommended changes in the way that costs are assessed by the federal government. "Although regulations continue to proliferate, the cost-benefit ratio has not been adequately assessed scientifically, ethically, or financially by society, legislators, and regulatory agencies," they said. The authors noted that "animal-based research comprises almost half the portfolio of the National Institutes of Health" and "is vital to the biomedical community but is encountering unprecedented challenges to its continued success." They cited increasing reliance on "designer rodents which require intensive health monitoring and more sophisticated husbandry" and the growth of molecular genetics, which "has brought many scientists with little or no background in animal research to animal-based studies." These scientists need to be trained to carry out animal research by veterinarians and other animal resource staff, a group which is also responsible for IACUC functioning and insuring institutions' compliance with federal regulations. "Animal welfare is a vital concern for the public and the scientific community; as a result, animal experimentation is heavily regulated. Regulatory agencies require extensive documentation of virtually all activities that involve animal use, entailing additional efforts from investigators, administrators, and animal resource staff," Cork and collegues pointed out. Under the circumstances, it is not surprising that laboratory staff, who already feel stretched to their limits, do not welcome continuing discussion on improvements in laboratory animal welfare. Any substantive changes will entail more money and more time spent monitoring compliance. This is not to condone resistance to improved standards of animal well-being, only to point out that such resistance is understandable when viewed from the researchers or the institution's perspective.

Imperfect as this process may seem, clearly efforts have been and are being made—by investigators, IACUC members, regulatory officials, and animal protection groups—to examine and refine the ways animals are used in research. This process is fraught with conflict and tension, but these negatives can be offset by good-faith efforts on the part of all players to do their best. For example, membership on an IACUC is time-consuming because the committee members are responsible for careful review and extended discussion of every protocol involving animals within the institution. In a large research facility, this can mean reviewing hundreds of protocols a year. If an IACUC rejects a research project, the investigator must then review the protocol and make the changes recommended by the group in order to submit it for funding. Typically, IACUC members serve on the committee for two years, and membership consequently revolves, with more experienced members serving alongside virtual novices.

The role of the unaffiliated member is particularly delicate. "When PHS [Public Health Service] policy and AWRs [animal welfare regulations] require appointment of an unaffiliated member, it is intended that the person should represent the interests of the public in the humane use of animals in research, testing, and education . . . in a pluralistic society 'general community interests' are increasingly difficult to identify," notes J. Wesley Robb, emeritus professor of religion and bioethics at the University of Southern California. Robb, who has served on two IACUCs, remarks that "the unaffiliated member is usually chosen from people who are likely to be cooperative with the review process . . . far too often, some benign individual is selected who may accept the appointment out of a sense of public service or because of a friendship within the institution . . . The letter of the obligation to appoint a person not affiliated with the institution is fulfilled, but the spirit of the rule is neglected." Robb argues that "the unaffiliated member can bring to the IACUC a fresh, different, and questioning perspective to the animal research enterprise."

On the most basic level, the unaffiliated member can insist that the investigator or members of the committee rephrase the information in the protocol in layperson's terms. The technical language of a protocol, Robb observes, "may obscure both the ethical issues and the science involved in the research proposal. Over the years, I have frequently had to inquire, 'What are you really doing with these animals? How invasive is the procedure?'" These types of questions are essential, he says, because "it is possible for the investigator to focus on the goals of the research effort and become inured to the stress or the pain the animal is experiencing. One of the functions of the committee is to review situations where this may occur."

Robb admits that his probing of these issues has not always been valued or appreciated by other members of the committee. "I have felt at times that my comments and concerns were considered by some members of the committee as a form of harassment having little or no substantive value. Unless this type of feeling is openly addressed by the 'outside member' with the committee, mutual respect and dialogue will be threatened. Lay members are not scientists; that is why they are on the committee. Lay members should not be treated in a patronizing way, nor should lay members have a chip on their shoulders; they should be equally respectful of the scientist's interests and professional expertise. Mutual understanding and respect is fundamental to fulfilling successfully the intent of the mandate that a community representative serve on the committee."

In some countries (notably Sweden and the Netherlands), individuals identified as supporters of animal welfare or animal rights serve on ethical review committees and act as advocates for experimental animals. This has not been the case in the United States, where those who are perceived (and perceive themselves) as advocates for animals have largely been prevented from serving on IACUCs. Opponents have argued that supporters of animal rights (and to a lesser extent, organized animal welfare) do not represent the viewpoints of the typical American and as such have no place on the committee. This perspective is given some support by the few studies of animal rights advocates that have been conducted to date.

In a paper titled "Ethical Ideology and Animal Rights Activism," psychologists Shelley Kaplan and Harold A. Herzog, Jr., note that "despite the impact of the animal rights movement on science, little is known about activists themselves. Stereotypes abound. They are portrayed variously as sentimental bleeding hearts, anti-intellectuals, terrorists, and misanthropes who love animals and loathe people. With few exceptions, however, there is a lack of research on the psychology of animal rights activism."

But on the basis of four years of research, including extended interviews with animal rights activists and responses to an "ethics position questionnaire," a test measuring "two dimensions of ethical ideology: relativism and idealism," administered to more than three hundred participants at the 1990 March for Animals, the authors felt able to draw some preliminary conclusions.

Most generally, Kaplan and Herzog found that the acceptance of animal rights ideology created profound changes in the lives of its adherents. An overwhelming majority (97 percent) had made changes in their diets, becoming vegetarian or vegan, and/or purchasing habits, with 94 percent stating that they bought only "cruelty free" products. Most refused to wear fur or leather and participated in various forms of protest, from writing to legislators to participating in rallies and marches to liberating research animals. "In some ways the shift in their thinking was similar to religious conversion," Kaplan and Herzog wrote. "It is our sense that many (not all) activists incorporate into their thinking a world view that is idealistic, evangelical, and distrustful of science and technology."

Kaplan and Herzog found that of the four ethics positions measured by their testing instrument (situationist, absolutist, subjectivist, and exceptionist), animal rights activists fell disproportionately into the absolutist category. "Absolutists (low relativism, high idealism) believe that there are universal moral principles and that adherence to them will almost always lead to positive consequences and the protection of the general welfare," as opposed to subjectivists (high relativism, low idealism), who "believe neither in universal moral principles nor that ethical behavior will always lead to positive outcomes." Noting that the testing instrument was originally used and validated among college students, Kaplan and Herzog found that "as predicted, a higher proportion of activists than students were classified as absolutists in their moral orientation. Conversely, the proportion of respondents classified as subjectivists was much lower among the activists than among the student sample."

The researchers concluded that "our results suggest that the majority of activists attending the March for Animals were committed to an ethical ideology that is idealistic and absolute. This finding, if characteristic of the movement as a whole, has several implications. First, absolutists may be less willing to compromise than individuals who view moral judgement through a more relativistic lens. In this context it is not surprising that scientists and activists have difficulty reaching common grounds for discussion. Second, it is possible that moral orientation is a factor that facilitates involvement in the movement for some and virtually precludes involvement for others; we were particularly struck by the near absence of subjectivists in our sample.

"While these results should be regarded as suggestive," they said, "they support the idea that members of the animal rights movement are marching to a different drummer. They live in a different moral universe than most Americans. Furthermore, many change their lives to bring behavior in line with belief."

Jasper and Nelkin, the New York University sociologists who described animal rights as "a moral crusade," noted that "the language of moral crusades is sometimes shrill, self-righteous, and uncompromising, for bedrock principles are non-negotiable . . . Their sense of moral urgency encourages believers to ignore laws and conventional political processes, and they organize themselves into groups structured for quick action, not participatory debate. Proselytizing and unconventional in their style, such crusades frequently appear dangerous to those who do not share their judgmental and uncompromising views."

This has certainly been the case with animal rights, particularly as it is viewed by the research community. Although the political power wielded by proponents of animal rights has been proportionately small and their influence on the conduct of biomedical research seemingly restricted to reforms supported by the majority of Americans, they have been perceived as powerful adversaries by researchers. "The animal liberation/rights movement (ALRM) poses a significant threat to the future of medical research. This threat is far greater than currently recognized by most physicians," Richard P. Vance wrote in the *Journal of the American Medical Association.* "We have consistently misjudged and underestimated the potential danger of the ALRM."

Unlike much of what is published by researchers critical of the animal rights movement, Vance's article does not dismiss the claims of animal protectionists as sheer emotionalism or sentimentality. Vance clearly makes an effort to address the philosophy behind animal rights, as expressed in the work of its two chief architects, Tom Regan and Peter Singer. "Both are exceptionally good philosophers in the analytical tradition. They provide sophisticated defenses of their positions," he says. Nonetheless, he finds defects in the tradition of analytical ethics they embody, in the "limited nature of the philosophical tools they use. Their ultimate theoretical weaknesses are extremely common among analytical ethicists. Unlike more substantive ethical traditions (for example religious or ethnic traditions), analytical ethics cannot draw on a rich array of sources—canonical texts, authoritative readings, overlapping (even contradictory) platitudes, interpretative communities, and the like. In comparison with such traditions, analytical ethics is abstract and thin. Despite claims of rational consistency, no analytical model has been able to claim adequacy."

Although the philosophy supporting animal rights is weak in his view, Vance does not counsel complacency. "We can no longer afford comfortable but erroneous myths that make the ALRM appear emotional, ignorant, and brainwashed. At its best, the ALRM is intelligent, politically astute, and resourceful," he says. Yet, in a conclusion that neatly prophesies the tension within the movement that was already beginning to appear in 1992 and would reach its conclusion in Gary Francione's *Rain Without Thunder* six years later, Vance notes that "there are deep philosophical divisions within the ALRM that vitiate their pretense of a unified movement. Recent philosophical efforts in the ALRM already seek ways to begin to overcome the weaknesses in Regan and Singer. More careful attention needs to be paid to these efforts. The future of medical research could be in jeopardy if we fail to understand our most sophisticated and dangerous critics."

By 1998, however, animal rights criticism of biomedical research in the United States had begun to seem anything but sophisticated. As the incidents described earlier in this chapter illustrate, the most visible critics of animal research appear to affirm the scientific community's assertion that the movement is the exclusive province of thugs and fanatics whose shrill denunciations of research are viewed with distaste by the majority of Americans. Like their nineteenth-century antivivisectionist forebears, these activists have alienated potential supporters with absolutist rhetoric and behavior that has greatly diminished the moral legitimacy of their cause in the eyes of the public. As a consequence, public support for criticism of scientific research based on animal rights claims has greatly atrophied in the United States.

However, a group of new technologies that necessarily involve extensive use of animals has the potential to reignite the debate. Xenotransplantation, in which organs and tissues from animal donors are transplanted into human hosts, and transgenic technology, which adds or deletes genes or transfers genetic material between species, are two types of research that have raised questions and concerns among some scientists and bioethicists. Even more controversy surrounds a related development, still in its infancy but already a source of considerable discussion: somatic cell nuclear transfer, popularly known as cloning. All of these use significant numbers of animals in research programs that deserve careful scrutiny. But most of the debate thus far, both within the scientific community and among bioethicists and policymakers, has focused on the possible implications of these evolving technologies for human beings, with relatively little discussion of their impact on animals or the linkages between the two.

"I find it unfortunate that people in this country pay so little attention to what is being done in the field of animal biotechnology," commented bioethicist Erik Parens in 1999. Following a talk given at the University of Maryland titled "Ethics and Genetics: From Testing Ourselves to Designing Our Children," Parens noted that although biotechnology is now being used to produce animals that are tailor made for various human purposes, this research, which is already quite advanced, has been the topic of very little discussion in the United States, while it is a major public policy issue in Europe. One reason for that may be that agricultural research on animals is not covered by the U.S. Animal Welfare Act, which regulates only biomedical research on animals. Then, too, farmers and animal breeders have for centuries bred animals to enhance or eliminate particular traits. The difference is that today these manipulations can be carried out on a genetic level and thus have profound implications for human medicine and for society. The same process of somatic cell nuclear transfer that was used by Ian Wilmut and his colleagues at Roslin Institute, the celebrated research facility in Scotland, to clone a sheep with certain desirable traits "could in the not too terribly distant future be used to produce human beings whose traits we value," Parens said, or to eliminate certain genes that cause suffering and death.

At this point, most responsible members of the scientific community reject the former possibility while embracing the latter. But many scientists do not see any moral or ethical obstacles in using a combination of somatic cell nuclear transfer and other biotechnologies to help infertile couples conceive a biologically related child, for example, or to treat genetic disease.

All these possibilities must be extensively researched and tested in animals before being applied to human populations, and this preliminary stage would appear to be a good time to discuss what limits, if any, society might choose to impose on this research and what uses of the new technology it is willing to permit. But the animal work is proceeding with minimal public debate in the U.S.—and, like the in vitro fertilization research that preceded it, no government regulation. The RAC (Recombinant DNA Advisory Committee) and NBAC (National Bioethics Advisory Committee), panels of scientists and bioethicists formed to examine such issues, have no authority to promulgate regulations or formulate policy and work in virtual isolation from each other, policymakers, and the public. The same is true of ELSI, the bioethics component of the Human Genome Project. The work of these committees often appears to take place in a vacuum, with little coverage in the media of their deliberations and seemingly little public interest in the topics under discussion except among those professionally concerned with the issues.

Although this may seem a comfortable state of affairs for researchers today, permitting them to explore the scientific possibilities of new technologies without burdensome regulations or interference, past experience indicates that when science surges too far ahead of public comprehension, an inevitable backlash ensues. A conversation on the issues raised by some of the new technologies and their relationship to long-standing tensions expressed in the research/animal protection conflict seems long overdue.

NINE

ENGINEERING LIFE

We know so little about the human body. We know so little about life itself, that we should not try to dedicate engineering to try to improve anything. What our society does fifty years from now is its business. It doesn't care what we think, and we don't care what people of fifty or a hundred years ago thought we should do. But it is our duty to go into the era of genetic engineering in as responsible a way as possible. And that means to use this powerful technology only for the treatment of disease and not for any other purpose.

—W. French Anderson, director,
Gene Therapy Laboratories, University of
Southern California School of Medicine

WASHINGTON, D.C., 1998

OUTSIDE, IN the near-tropical heat of a Washington summer, tourists roam up and down Pennsylvania Avenue. Inside a hotel a few blocks from the White House, a small group of scientists, philosophers, entrepreneurs, and ethicists is meeting to discuss a new development in science that the tourists and their fellow citizens fear and distrust. At least, that's what the polls say. Worse yet, under the majestic dome of the Capitol a few blocks away, lawmakers have discussed banning the new development. This possibility fills the scientists with dismay. How can they persuade the public and Congress that there is nothing to fear? That scientists know what they are doing and won't go too far? That the benefits far outweigh the risks?

This is an old story become new once again. One hundred and fifty years ago people feared vaccination. Today they fear cloning. The very name raises terrible visions in the public mind, attendees at the Second Annual Congress on Mammalian Cloning agree. "When the public hears the word *cloning*, they think of Frankenstein," says Lee M. Silver, a professor of molecular biology at Princeton and one of the cochairs of the meeting. "I recommend that we not use the word, to reduce public fear. Let's use 'somatic cell nuclear transfer' or 'somatic cell genetic transfer.' 'Organic enhancement' is even better. People like that. But I'm not sure it's going to work."

Indeed, public anxiety surrounding cloning is well documented and has already resulted in anti-cloning legislation in California. In March 1997, Representative Vernon Ehlers introduced a bill in the House of Representatives (H.R. 922) to ban human cloning in the United States, and hearings were held, though no legislation has been passed. In Europe, legislators and the public are even more restive. In January 1998, President Jacques Chirac of France proposed banning human cloning throughout Europe via a European Convention on Human Rights and Biomedicine. Nine months later, most EU countries had signed the Council of Europe protocol banning human cloning, although the British have refused because they believe the legislation is too restrictive, and the Germans, with the memory of 1933–45 still sharp, believe that it is not restrictive enough.

In Switzerland, anti-cloning sentiment is bleeding into related areas of biotechnology. For example, the Swiss only recently approved continued research on "transgenic" animals—animals transfected with human genes or the genes of other species. By 1998, transgenics had become one of the most promising tools of biomedical and pharmaceutical research, with custom-made animals genetically altered to serve as models for a broad range of human diseases. Transgenic animals are also being used as "bioreactors," with genetic modifications that enable them to produce valuable human proteins and pharmaceutical products in their milk—a practice known as pharming. The Initiative for the

Protection of Life and the Environment against Genetic Manipulation would have banned the patenting, production, sale, and acquisition of genetically modified animals, and outlawed field tests of genetically altered plants. If the Swiss had approved the nationwide referendum to outlaw transgenics, the pharmaceutical and biomedical research industries in Switzerland would have been decimated. In the end, many observers believe that it was the economic issue that convinced Swiss voters to oppose the referendum by a margin of two to one. The Swiss, after all, are a practical people.

Genetic engineering of plants has also led to agitation and resistance in other European countries. "Last summer, biotech food emerged as the most explosive environmental issue in Europe. Protestors have destroyed dozens of field trials of the very same 'frankenplants' (as they are sometimes called) that we Americans are already serving for dinner, and throughout Europe the public has demanded that biotech food be labeled in the market," Michael Pollan wrote in 1998 in *The New York Times Magazine*. "Austria, Luxembourg and Norway, risking trade war with the United States, have refused to accept imports of genetically altered crops. Activists in England have been staging sit-ins and 'decontaminations' in biotech test fields."

Public agitation over this issue continued into 1999 and spread to the United States, although U.S. protests have not been as widespread as those in Europe. Nonetheless, the potential political impact of the issue has caused companies like Monsanto, heavily invested in the genetic engineering of plants, to launch expensive public relations campaigns to try to convince consumers that their products are safe for people and the environment. As of January 2000, such campaigns had not been particularly successful. In fact, according to an article by Justin Gillis published in *The Washington Post* on October 26, 1999, Monsanto stock had lost more than a third of its value because of the international uproar over the safety of bioengineered seed and food products. Meanwhile, Europeans continued to decry what they perceived as American indifference to the dangers posed by genetic engineering; an article published by the *Guardian* on March 2, 1999, congratulated Britons for "resisting the combination of wild-eyed techno-utopianism and stock market–fueled greed that, together with incessant lobbying by the genetic-industrial complex, has effectively stifled debate on genetic engineering in the United States."

The American scientists and entrepreneurs at the 1998 cloning congress do not want the antibiotech hysteria raging in Europe to spread to the United States. Like the Swiss, Americans are known to be pragmatic (when not irrationally ideological), which makes the proponents of biotechnology at the meeting hopeful that once the American public has been educated about the benefits of some of the new technologies, particularly somatic cell nuclear transfer (cloning), they will become excited about the possibilities. What if inherited diseases like Tay-Sachs, sickle-cell anemia, and certain kinds of cancer could be eradicated? What if infertile parents who desperately want a biological child could have one using nuclear transfer technologies? What if cloning could be used to create custom-made organs for those needing transplants or genetically alter animal organs and tissues to prevent their rejection by human hosts? What if parents could actually use a combination of in vitro fertilization and cloning, together with some of the other new developments, such as the artificial chromosome, to actually program their offspring, enhancing their intelligence or athletic ability, deciding the color of their hair and eyes?

This last group of what-if's has some of the scientists and ethicists who were at the meeting worried. Eliminating disease is one thing, but genetically programming offspring for desired traits is another. Still, much has changed in the year that has elapsed since the First Annual Congress on Mammalian Cloning, held in June 1997. The first congress was held in the immediate aftermath of the international uproar over Dolly, the sheep that Ian Wilmut, Keith Campbell, and colleagues at their Roslin Institute claimed as the world's first true animal clone, produced by fusing an enucleated donor egg with the genetic material of an adult cell. Newspapers and magazines around the world broadcast the sensational news. Ian Wilmut, the mild-mannered fifty-two-year-old British scientist who headed the Roslin research team, was catapulted overnight into the ranks of international celebrity. Two days after Dolly's creation was announced, the president of the United States instructed the National Bioethics Advisory Committee to produce a report in ninety days on the ethical implications of Wilmut's discovery and the possibility of human cloning. Concluding that ninety days was hardly enough time to resolve the ethical issues surrounding cloning, the committee said, that a 1994 ban on the use of federal funds for creating human embryos for research purposes applied to cloning as well, although critics argue that this leaves the door open for privately financed clinics and laboratories to conduct such research.

One year later, some scientists still have trouble believing that Wilmut and his colleagues were actually successful in reducing an adult cell to a dedifferentiated state, reversing the cell's developmental process to the point that when its genetic material was extracted and fused with an enucleated donor egg, the process of cell division proceeded as if the organism were wholly new. The scientific implications of that act, never mind the ethics, are mind-boggling. "I still have trouble believing that Dolly is Dolly, that an adult fully differentiated cell can dedifferentiate back to ground zero," said French Anderson, cochair of both the first and second cloning congresses, in his opening remarks at the second congress.

Anderson is director of the gene therapy laboratories at the University of Southern California School of Medicine. A clinician, he sounds excited by the medical applications of somatic cell nuclear transfer, envisioning the therapeutic relief that could be provided if scientists and doctors really understood and could control the process of cell formation, differentiation, aging, and death—if they could reprogram an aged cell back to virtual infancy, and reprogram or replace a damaged cell. "The implications are profound. If we can reset all those switches, then we can learn to reset them anywhere along the line. Cancer, aging, degenerative diseases—the medical implications are profound. So it's extremely important to know if it's really true," he says.

Like most of the scientists attending the first congress, in June 1997, Anderson admits that at the start of the meeting he was excited by the science, but shaken by possible human applications of the technology. By the end of the congress, however, after two days of animated discussion with other leaders in the field, he found that his perspective had changed. "We took a straw poll in the aftermath of the first conference," Anderson commented in his introductory remarks at the second congress, in 1998. "Before that meeting, our feeling about the application of this new technology to humans was very negative. But over the course of a couple of days it became clear that the issues were not as simple as we thought. We went from an emotional feeling to one of more understanding. The biggest concern that most of us have now is that there be no legislation banning human cloning." Any legislation seeking to regulate cloning must, he says, distinguish "what is valuable and important from what is repugnant" and that "the only really repugnant application would be rich egocentric people who wanted to clone themselves."

If the scientists and ethicists at the 1997 congress were reeling (like everyone else at the time) from the philosophical and emotional impact of Dolly, one year later many appeared drunk on the possibilities and the power accruing from that achievement. Some, like Princeton molecular biologist Lee M. Silver and philosopher George Ennenga, speak of the potential to direct the course of human evolution. Others, more restrained, discuss the benefits of cloning technologies for agriculture and medicine. But until the final day of the meeting, no one mentions animal welfare. Although Ian Wilmut and his colleagues at Roslin Institute are sensitive to welfare issues and operate under the stringent British Animals Scientific Procedures Act, the technique used to produce Dolly inarguably uses large numbers of animals (277 pregnancies failed before the successful birth of Dolly), with most offspring prone to defects that result in the animals' deaths in utero or soon after birth. As bioethicist Arthur Caplan observed in a 1997 article in *Business Week* heralding the new discovery, "They killed a lot of embryos and made a lot of malformed sheep."

Tanja Dominko, a primate researcher at the Oregon Regional Primate Center, comments in her presentation that her group must proceed very carefully with the technology because "our research would be shut down by animal rights activists" if primate researchers lost the number of animals agricultural researchers had. Otto Postma, CEO of Pharming Health Care Products, a subsidiary of a Netherlands-based biotech company, discusses the Dutch Animal Health and Welfare Act of 1992, which prohibits animal biotechnology acts unless the acts create no disproportionate animal suffering and no ethical objections. Questioned by one of the American scientists about the nature of such ethical objections, Postma answers that the procedures must not violate the integrity of the animal, the "cowness of the cow," for example. The American scientists are baffled by this concept. In the United States, such talk is considered the province of the most extreme animal rights proponents. In Europe, however, animal protection viewpoints considered extreme in the United States have been extensively discussed and enshrined in national and EU legislation. As Postma points out, the Netherlands Biotechnology Act and its animal welfare provisions are "the result of over ten years of discussions in the Netherlands" about science and animal protection issues.

But that kind of restrictive legislation is not what most of the meeting's attendees want to see in the United States. At the second cloning congress, and during the day-and-a-half meeting that preceded it, Transgenics and Cloning: Commercial Opportunities, the word used most often to characterize the type of communication necessary is *education*. The public needs to be educated on the issues in order to make an informed decision, speaker after speaker noted. Little doubt was expressed by the meeting's attendees that if the public were properly educated about the benefits of the technology, they would support further research on cloning, even that using human cells. As DNA pioneer James D. Watson pointed out at a symposium on germline engineering held a few months earlier in Los Angeles, "I think we can talk principles forever, but what the public wants is not to be sick. And if we help them not to be sick, they'll be on our side."

THE CLONING STORY began in 1952 with Watson himself, his colleague Francis Crick, and two feuding researchers at King's College in London. As Watson himself tells the tale in his celebrated but controversial account of the discovery of the structure of DNA, *The Double Helix*, he and Crick, working at the Cavendish Lab in Cambridge, consulted regularly with Maurice Wilkins, a physicist turned biologist like Crick, to discuss Wilkins's research and that of his colleague Rosalind Franklin at King's College. Unlike Watson and Crick, who were using models to theorize a possible structure for DNA, believed by some scientists to be the basic mechanism of heredity, Franklin and Wilkins were using a laboratory technique called crystallography to illuminate the structure of the DNA molecule. X-ray crystallography, a technique invented by William and Lawrence Bragg in 1913, creates an image on photographic paper, a three-dimensional representation of large biological molecules such as proteins and nucleic acids. Franklin's X-ray photographs clearly illuminated the helical shape of the DNA molecule. "The instant I saw the picture, my mouth fell open and my heart began to race," Watson testified in *The Double Helix*.

Franklin and Wilkins's research enabled Watson and Crick to refine their model building and focus on the problem of how the bases were held together. In 1953, Watson and Crick published one of the most famous scientific papers of all time in the journal *Nature*, announcing their discovery that DNA's shape was that of a "double helix," or a spiral ladder, with the four bases—adenine, thymine, guanine, and cytosine—held together in pairs (A-T, G-C) by hydrogen bonding. Zipped together, each strand of the helix, each side of the "ladder," contains all the necessary information to make a complementary strand of DNA when "unzipped." "It has not escaped our notice that the specific pairing we have postulated immediately suggests a possible copying mechanism for the genetic material," the two pointed out. Once Watson and Crick had nailed the structure, the stage was set for the creation of a new science, molecular biology. Beginning in 1961, scientists began to explore the genetic code, studying the actual molecular operations that dictated the process of Mendelian genetics, described by Gregor Mendel over one hundred years before. The actual mechanisms of heredity were slowly becoming clear. By 1990, researchers were hard at work identifying and sequencing the genetic codes of a wide variety of organisms, from microbes to mice to the grandest project of all—a multinational effort to map and sequence the human genome that, by June 2000, was deemed 90 percent complete.

"NATURE IS A NASTY BITCH," molecular biologist Lee Silver said at the Second Annual Congress on Mammalian Cloning, discussing the horrors of Tay-Sachs disease, a genetically transmitted condition prevalent among Jews of European ancestry. Tay-Sachs is a fatal disorder of the nervous system, leading to mental retardation, paralysis, and blindness. Symptoms appear early, when a child is less than six months old, and most of its victims die young, before age six. The disease is caused by a defective gene that prevents cells from producing enough of an enzyme called hexosaminidase A. When two carriers of Tay-Sachs marry and have children, tragedy often results. Techniques of DNA analysis developed in the 1980s have enabled prospective parents to determine whether or not they are Tay-Sachs carriers and to abort pregnancies where the fetus is shown to carry two copies of the defective gene. The number of babies born with Tay-Sachs disease in the United States has rapidly declined to about a dozen a year, based on prenatal screening and postdiagnostic abortion.

Using Tay-Sachs as an example of the suffering caused by genetic disease, it is easy to understand the hostility expressed by Lee Silver and difficult not to applaud the efforts of scientists to counter Nature's more egregious cruelties. Silver is not the first scientist to paint Nature as a villain and to characterize human efforts to control Nature as heroic. As infectious diseases have been tamed in affluent industrialized nations, genetic diseases have assumed much greater significance. The human desire to master the genome, to identify the causes of genetic illness, is thus merely the latest human attempt in a long struggle to shape the world according to human ends, needs, and desires. Genetic engineering of plants and animals, according to this view, is no different from earlier forms of selective breeding and cross-fertilization. Attempts to identify the causes and develop treatments for genetic diseases are no different from a search for vaccines against viruses or the discovery of antibiotics to treat bacterial infection. Our tools are just different now, and our understanding greater. But perhaps our understanding is not great enough, some say. Even among those committed to the scientific worldview, anxiety and fear sometimes take hold.

In 1971, Stanley Cohen at Stanford University and Herbert Boyer at the University of California at San Francisco created the first "recombinant" DNA, splicing bacterial DNA into the DNA from an animal virus to produce "chimeric" DNA formed from the genetic material of two different sources. Scientists working elsewhere soon discovered how to multiply the chimeric DNA by introducing it into *E. coli* bacterial cells, which multiply rapidly. At a summer conference in 1971 in Cold Spring Harbor, New York, Janet Mertz, an associate of researcher Paul Berg of Stanford University, described a planned experiment in which a tumor virus called SV40 would be inserted into a laboratory strain of *E. coli*. A number of researchers present at the meeting contacted Berg immediately afterward and requested that he refrain from carrying out this experiment, because of the possibility of creating an easily transmissible cancer-causing virus, which might escape from the lab and wreck havoc among the population. Berg agreed to postpone the experiment. "The Berg experiments scare the pants off a lot of people, including him," Wallace Rowe of the National Institute of Allergy and Infectious Diseases told science writer Nicholas Wade.

Still, as work on recombinant DNA continued in laboratories around the world, scientists grew increasingly uneasy. In June 1973, the cochairs of a conference on recombinant DNA research (Gordon Research Conference on Nucleic Acids, June 11–15, 1973) wrote to the president of the National Academy of Sciences, Philip Handler, to express their "deep concern" over the dangers inherent in certain types of recombinant DNA research. In a letter signed by a majority of the participants at the Gordon conference and reprinted in prestigious journals, including *Science*, Maxine Singer and Dieter Soll wrote that "scientific developments over the past two years make it both reasonable and convenient to generate overlapping sequence homologies at the termini of different DNA molecules. . . . This technique could be used, for example, to combine DNA from animal viruses with bacterial DNA, or DNAs of different viral origins might be so joined. In this way, new kinds of hybrid plasmas or viruses, with biological activity of unpredictable nature, may eventually be created. These experiments offer exciting and interesting potential both for advancing knowledge of fundamental biological processes and for alleviation of human health problems. Certain such hybrid molecules may prove hazardous to laboratory workers and to the public. Although no hazard has yet been considered, prudence suggests that the potential hazard be seriously considered."

The Singer-Soll letter led to the creation of the blue-ribbon Committee on Recombinant DNA Molecules of the National Academy of Sciences, headed by the same Paul Berg whose research had created such a ruckus at the Cold Spring Harbor meeting two years earlier. The committee produced the famous

Moratorium letter, published in *Science* on July 26, 1974, advising a voluntary cessation of recombinant DNA research until some attempt had been made to "evaluate the hazards and some resolution of the outstanding questions has been achieved." This committee, whose members included current and future Nobel Prize winners Berg, David Baltimore, James Watson, Stanley Cohen, Herbert Boyer, and Daniel Nathans, did something unprecedented in the history of science. It recommended that researchers throughout the world put aside professional competition, the desire for renown, and the thirst for knowledge and take time to consider the implications of their work. This had never happened before, and hasn't happened since. Even when Dolly appeared in 1997, nothing comparable to the Berg letter, with its august list of signatories, appeared counseling a moratorium on further research until costs and benefits could be assessed. Why did it happen in 1973?

In the prologue to *The DNA Story: A Documentary History of Gene Cloning*, James D. Watson and John Tooze partially answer that question: "Increasingly during 1973 we began to ask whether in the process of possibly discovering the power of 'unlimited good' we might simultaneously be setting the power of 'unlimited bad.' Might some of the new genetic combinations that we would create in the test-tube rise up like the genie from Aladdin's lamp and multiply without control, eventually replacing preexisting plants and animals, if not man himself? . . . from the start we knew that no one had concrete facts by which to gauge these scenarios of possible doom. So perhaps we had best proceed in the fashion of the past 500 years of Western civilization, striking ahead and only pulling back if we find the savages not of normal size but of the King Kong variety, against which we have no chance?"

But, the authors note, "this was not the mood of the early 1970's when many academics thought that science was already out of control." In the wake of the Vietnam War, civil rights struggles, and a general distrust of all kinds of cultural authority, even among scientists themselves, a different approach was needed. "For us to move forward full steam ahead with recombinant DNA without considering deeply what havoc it might wreak obviously seemed socially irresponsible." Nonetheless, as Watson and Tooze bluntly admit, "it was never the intention of those who might be called the Molecular Biology Establishment to take this issue to the general public to decide. . . . We did not want our experiments to be blocked by overconfident lawyers, much less by self-appointed bioethicists with no inherent knowledge of, or interest in our work. Their decisions could only be arbitrary. . . . Recombinant DNA as a social issue thus started as a dialogue between scientists."

It did not stay that way. Molecular biologists met for three days at Asilomar, a beautiful conference center in the woods on the shores of the Pacific about two hours south of San Francisco, in February 1975. The meeting's purpose was to "review the progress, opportunities, potential dangers and possible remedies associated with the construction and introduction of new recombinant DNA molecules into living cells." The International Conference on Recombinant DNA Molecules, attended by 140 scientists, led to a set of NIH guidelines advocating stringent containment procedures in laboratories working with recombinant DNA. Experiments were to be classified according to risk (P1–P4), with increasingly rigorous requirements as the possible dangers increased. Certain experiments were to be deferred, including "the cloning of recombinant DNA's derived from highly pathogenic organisms, DNA containing toxin genes, and large-scale experiments using recombinant DNA's that are able to make products potentially harmful to man, animals, or plants."

Most molecular biologists were satisfied. They had met, discussed, and come up with a set of workable guidelines. Some might argue that the guidelines were too strict, and others, that they were not strict enough. Nonetheless, the scientists believed that they had confronted the issues in a responsible fashion and that work could proceed. The public (and certain high-profile researchers) disagreed. Before the Asilomar conference, relatively few members of the general public were aware of recombinant DNA research and its implications. After Asilomar, it was impossible not to be aware of both the possible benefits and the shadowy dangers that scientists themselves admittedly feared. Furthermore, the fact that scientists believed that they could simply decide these matters among themselves, without consulting the public, disturbed many people. The Genetic Engineering Group of Science for the People, a group of researchers and scientists opposed to the characteristically insular style of decision making exhibited at Asilomar, said in an open letter to the attendees of the conference, "Since the risks and dangers of these technologies are borne by the society at large, and not just the scientists, the general public must be directly involved in the decision making process. Yet we see that even in the structure of this conference that a scientific elite is here alone trying to determine the direction that such regulation should take."

In 1976, the controversy erupted in Cambridge, Massachusetts, when Harvard proposed to convert some rooms in the biology department into a P3 (moderately high level containment) laboratory devoted to recombinant DNA research. Scientists with offices in the building protested vehemently and took their concerns to the mayor, Alfred Velucci. In the summer of 1976, the City Council held hearings and heard testimony from those in favor of and opposed to construction of the labs, which included Nobel Prize–winning biologist George Wald and his wife and fellow biologist, Ruth Hubbard. The City Council voted on a three-month moratorium on recombinant DNA research in the city of Boston, a ban that would affect both Harvard and MIT, the city's premier research institutions. The council also set up a Cambridge Experimentation Review Board to draft recommendations on recombinant DNA research that were expected to be much stricter than those set by NIH after the Asilomar conference.

While the battle raged in Cambridge, similar discussions were being held all over the country and, more disturbing to some, in the pages of respected science journals. Unlike the Asilomar group, who felt that the primary risks were related to safety and that the guidelines had addressed those issues, the public and a few key scientists were just as disturbed by the long-term biological and philosophical implications of the new science of genetic engineering, issues that were not touched on at Asilomar. Erwin Chargaff—a renowned genetics researcher whose finding that in the DNA of any species the number of adenine bases equals the number of thymine bases, and the number of cytosine bases equals the number of guanine bases, established a foundation for Watson and Crick's discovery of the structure of DNA—spoke out bluntly about the dangers inherent in the new research. At a public hearing on recombinant DNA research in New York, Chargaff said that "there is some importance in genetic experimentation and I have been too long a scientist to say certain kinds of experimentation should be forbidden by the authorities. But they should be controlled and I would not leave it to the scientists themselves to control themselves to ensure safe operations. Who will watch the watchers?" One year later, in an article titled "Playing God with DNA," Chargaff remarked, "It was quite clear that as soon as science was almost completely funded by the government, attempts would be made to push it into so-called useful channels. Into therapeutic uses for genetic diseases, a form of eugenics. I am one of the few people who remember that the extermination camps in Germany began as experiments in eugenics. So I am almost congenitally opposed to improving the human lot. It begins with the do-gooders and ends with the exterminators."

Liebe Cavalieri, another esteemed scientist, testified at the New York hearing (and in a related article in *The New York Times Magazine* published on August 22, 1976) that "the evaluation of benefits and risks and the guiding rules of the game have been made by scientists, with only token input by nonscientists. Is it not the public to whom the results, good or bad, will accrue? It is presumptuous of the scientific community to assume that its interests coincide with those of society as a whole. Furthermore, in acting on this assumption, they have preempted a public function, that of decision making where its own fate is concerned. Scientists must obviously play a role in this process, but not the only role."

Finally, Nobel laureate George Wald, who had played an instrumental role in the Cambridge incident, wrote the following in *The Sciences* in an article titled "The Case Against Genetic Engineering": "Up to now living organisms have evolved very slowly, and new forms have had plenty of time to settle in. It has taken from four to 20 million years for a single mutation, for example the change of one amino acid in the sequence of hemoglobin or cytochrome c, to establish itself as the species norm. Now whole proteins will be transposed overnight into wholly new associations, with consequences no one can foretell, either for the host organisms or their neighbors. It is all too big, and is happening too fast. So this, the central problem, remains almost unconsidered. It presents probably the largest ethical problem that science has ever had to face."

Wald ended the article by saying "my feelings are ambivalent, for the new technology excites me for its sheer virtuosity and its intellectual and practical potentialities; yet the price is high, perhaps too high. We are at the threshold of a great decision with large and permanent consequences. It needs increasing public attention here and worldwide, for it concerns all humankind. That will take time, during which we can try to learn, as safely as that can be managed, more of what to expect, of good and ill. Fortunately, there is no real hurry. Let us try, with good will and responsibility, to work it out."

The research moved forward with astounding rapidity. The original NIH guidelines, published in 1976, covered DNA research on microorganisms. By the time the NIH began developing appendices to cover research involving plants and animals, in 1986, the U.S. Supreme Court had ruled that recombinant microorganisms could be patented (*Diamond v. Chakrabarty*, 1980); the first recombinant DNA pharmaceutical, insulin, had been approved for sale in the United States (1982); the first cross-species transfer of a human gene into another mammal had been successfully executed (1982); and the first transgenic plant created (1982). By the time the revised NIH guidelines were approved in 1994, hundreds of different varieties of transgenically altered mice had been created and were being used in research laboratories around the world, genetically engineered plants were being sold in supermarkets, and Ian Wilmut and his colleagues were beginning the experiments that would lead three years later to the birth of Dolly.

By the time the Second Annual Congress on Mammalian Cloning was held in June 1998, biotechnology had become big business and the Human Genome Project was nearly three-quarters finished. The fears about the danger to public health and safety expressed by the pioneers of genetic engineering and their critics had apparently not materialized. No mutant virus had escaped from a lab to menace the earth and its inhabitants. But the philosophical issues loomed larger than ever, linked to concerns about related advances in biomedical research, particularly xenotransplantation.

IN FEBRUARY 1996, when scientists at San Francisco General Hospital transplanted baboon bone marrow cells into AIDS patient Jeff Getty in an effort to jump-start his decimated immune system, animal rights protesters argued that using animals as spare parts for human beings was morally wrong. But by then, animal rights protests against research were viewed with increasing suspicion by the public. Most Americans were aware that the rules had been tightened up since the Silver Spring monkey and University of Pennsylvania Head Injury Clinic incidents—and weren't most cosmetics and personal care products now "cruelty free"? The experimental use of animals no longer agitated large segments of the public, although incidents like the May 1997 revelation by a PETA undercover investigator of animal abuse at a contract testing laboratory in New Jersey still made news. And the need for organs for various types of transplantation was inarguable—approximately 48,000 people in the United States alone are waiting for an organ, with only about 7,500 Americans a year donating organs after their death. In the United Kingdom in 1997 approximately 5,000 people were waiting for a kidney, with an even smaller population to provide organs for donation. About 50 percent of people waiting for an organ die before one becomes available.

Nonetheless, reservations about the practice of xenotransplantation began to spread among both activists and virologists. The latter group focused not on the ethics of the procedure but on its practical dangers. What if a lethal primate virus, something like AIDS, were transferred into a human host by accident through xenotransplantation? What if the virus mutated in its new host, as viruses tend to do, and assumed an airborne form? In an echo of the recombinant DNA controversy of the 1970s, voices were heard within the scientific community asking for a moratorium on further xenotransplantation research until the risks could be properly assessed and safety regulations put in place.

Reports were issued by advisory groups to the British government and the United States Institute of Medicine in April and July 1996, respectively. Both supported further xenotransplantation research with a few caveats. The Nuffield Council, the U.K. bioethics group, advised against any further xenotransplantation research with primates for both safety and ethical reasons, noting both the dangers of cross-species virus transmission and the degree of similarity between humans and primates. The U.S. Institute of Medicine report recommended the development of national guidelines for further research and the creation of an advisory committee to coordinate (although not to regulate) further transplantation research. The Institute of Medicine also suggested that recipients of xenografts should be monitored throughout their lives through the creation of a national registry of xenotransplant patients, with "periodic surveillance" of friends and family members, to provide an early warning system for any viral contagion. In September 1996, the U.S. Food and Drug

Administration, the Centers for Disease Control and Prevention, and the National Institutes of Health released draft guidelines for xenotransplant research incorporating these recommendations. The guidelines stated that animals used in xenotransplant research "should be procured from screened, closed herds that arc well-characterized and as free as possible from infectious agents."

Although both the U.S. and British committees cautiously gave a green light to continued xenotransplantation research on animals bred in pathogen-free environments, with the British committee focusing on pigs as a less risky source of organs, two years after the reports were published, questions about pig viruses began spreading in the scientific community. In January 1997, a British government panel approved continued research on the transplantation of pig organs into humans, but also "called for more research into the potential for new human diseases arising from exposure to pig viruses." Author Nigel Williams added that "virologist Robin Weiss at the Institute for Cancer Research in London has shown in culture experiments that a retrovirus— which is carried in the genome and can be passed from parent to offspring—found in pig cells could infect human cells. This raises fears that retroviruses may infect patients receiving xeno-transplants and even spread more widely among the population." Meanwhile, animal protectionists argued that xenotransplantation was not only unethical but also a misallocation of tight health care dollars and a danger to public health. The animal protectionists argued for mandatory organ donation programs to meet the shortage of human organs for transplant.

At the mammalian cloning congress in June 1998, participants discussed the impact of cloning technology for xenotransplantation research. Steven L. Stice, chief scientific officer at Advanced Cell Technology, a Worcester, Massachusetts, biotech company, summarized an article that he and colleagues had published a few months earlier in *Nature Medicine* on the use of cloned transgenic fetal pig cells to treat Parkinson's disease. Parkinson's disease is a neurological disorder that usually affects people aged fifty to seventy. Symptoms develop gradually, with an initial trembling in a hand leading to an increased loss of muscle control and lack of balance. The disease is caused by the destruction of nerve cells in the brain that produce dopamine, a chemical important in cell communication.

By 1990, scientists had discovered that implanting fetal cells in the brains of Parkinson's patients alleviated some of the symptoms of the disease. However, the supply of human fetal tissue is limited, and many Americans oppose its use in biomedical research. In 1994, based on the recommendations of an advisory panel, Congress banned research on human embryos in publicly funded research. "Embryo research sits smack-dab in the midst of abortion politics," as bioethicist Arthur Caplan said at the 1997 cloning meeting, and there is no more contentious issue in U.S. political life than abortion. In Europe, abortion is not a political issue, so embryo research has not been affected. For example, in the United Kingdom, it is possible to fertilize a human egg and conduct research on the rapidly growing embryo for up to fourteen days. But in the United States, research using human embryonic tissue cannot be conducted using public funds, although privately funded laboratories, including in vitro fertilization clinics, are not prohibited from carrying out such investigations. Although many researchers privately say that science is being held hostage to the antiabortion lobby and that it would be far better for such research to be publicly funded and thus subject to regulatory scrutiny, few scientists in the United States have cared to argue publicly for their right to use embryonic cells in research. Antiabortion activists are even more vocal (and in many cases more violent) than animal rights activists. So U.S. researchers began using embryonic dopamine-producing cells from pigs and mice in a rat model of Parkinson's disease.

One of the greatest problems in any type of xenotransplantation is that the host almost immediately rejects the foreign tissue or organs. A rat's immune system recognizes that mouse cells are alien invaders and kicks them out like an unwelcome houseguest. Researchers discovered that if they suppressed the immune system of the rat, both pig and mouse cells would survive in the rat for a time. By using transgenic technology, it might be possible to transfer rat genes into the mouse before xenotransplantation, to "trick" the rat's body into accepting the foreign tissue indefinitely. If the genetically altered cells could be cloned, researchers would have succeeded in creating an endless supply of dopamine-producing cells for xenotransplantation. As Stice explained at the cloning meeting, he and his colleagues took this research one step further by investigating the possibility of using cloned fetal tissue from a transgenic cow. In the *Nature Medicine* paper, they concluded that "in order to make neurotransplantation a widely available and predictable therapy for Parkinson's disease, a uniform cell source is necessary. Cloned animal embryos offer the possibility of an abundant source. Cloned animals make it possible to genetically engineer cell surface antigens and insert other modulators of the immune response." With a nod to those who fear the possibility of cross-species viruses, the researchers added that screening of the clones for viruses could improve safety.

"Cloning is a faster and less expensive method of making transgenic animals," he concluded. Like a number of other researchers, Stice and his colleagues are also working on the xenotransplant animal of choice, the pig, attempting to create porcine transgenic clones for xenografts of solid organs into humans. Pig hearts are about the same size as the human heart, but like the mouse fetal dopamine cells rejected by the rat host, a pig heart will immediately be rejected by a human host, unless transgenic technology is able to engineer a pig heart with human proteins and antigens.

Eighteen months after the Second Annual Congress on Mammalian Cloning, scientists at the Oregon Health Sciences University announced that they had successfully cloned a primate. The monkey clone, Tetra, was one of four embryos created by splitting a single eight-celled monkey embryo. The quadruplets produced were then implanted into two surrogate monkey mothers. While two fetuses developed as a result of this procedure (one in each mother), only one, Tetra, survived.

The embryo-splitting procedure had previously been used to create genetically identical farm animals and mice, but Tetra is the first primate brought to birth by this method, which differs from the somatic cell nuclear transfer technique that produced Dolly. The authors of the *Science* paper announcing Tetra's birth noted that the many fetal deaths and miscarriages caused by somatic cell nuclear transfer, together with evidence that the method produces genetically "old" animals, led them to adopt the embryo-splitting strategy. They admitted, however, that the technique was able to produce only four pregnancies in thirteen attempts. Newspaper accounts of the Oregon researchers' success stressed the potential human health benefits of the experiment, with some observers pointing out that scientists' ability to create genetically similar primates could help in the development of an AIDS vaccine.

IF THE ISSUE of animal welfare has largely been ignored in public discussions about cloning and related biotechnologies, the potential impact of this work on human beings has been extensively discussed. However, these discussions for the most part have centered on the emotional and philosophical implications of the new technologies, rather than on actual physical risks of the procedures—which most scientists assume will be overcome through extensive animal research before the new technologies are ever used on human beings. But the philosophical issues loom large, even among those committed to moving forward with the research.

"We're unraveling our own blueprint and beginning to tinker with it, which is extraordinary. It means that we are becoming subject to the same powerful forces of conscious design that are completely reshaping the world around us," commented Gregory Stock, director of the Science, Technology, and Society Program at UCLA's Center for the Study of Evolution and the Origin of Life, at the June 1998 cloning meeting. Stock, coorganizer of the public symposium "Engineering the Human Germline," held at UCLA in March 1998, is an unapologetic proponent of this process of "conscious evolution."

Others are not so sanguine about either the biological or the philosophical issues raised by this undertaking. As far back as 1975, some scientists saw both the possibilities inherent in biotechnology and the dangers it posed. In an article titled "Troubled Dawn for Genetic Engineering," Robert Sinsheimer, chairman of the division of biology at the California Institute of Technology, asked a series of thoughtful questions that have assumed an even greater urgency as the powers that remained hypothetical in 1975 have become increasingly real: "How far will we want to develop genetic engineering? Do we want to assume the basic responsibility for life on this planet—to develop new living forms for our own purpose? Shall we take into our own hands our own future evolution? . . . Can we really forecast the consequence for mankind, for human society, of any major change in the human gene pool?" Sinsheimer pointed out that "to introduce a sudden major discontinuity in the human gene pool might well create a major mismatch between our social order and our individual capacities. Even a minor perturbation such as a marked change in the sex ratio from its present near equality could shake our social structures—or consider the impact of a major change in the human life span. Can we really predict the results of such a perturbation?"

Many contemporary researchers, including gene therapy pioneer French Anderson, who applied for permission to carry out the first human germline gene therapy experiments in 1999, are willing to use developing biotechnologies for therapeutic ends but are resistant to so-called enhancement uses. "The fundamental point I'm trying to make is that we don't know enough about what the consequences would be, from a medical point of view, to attempt anything at this point but the treatment of serious disease," Anderson said at the March 1998 symposium on human germline engineering. "The normal aging process I would not consider disease. The consequences of aging, namely cancer, heart disease, stroke—those degenerative processes that take place—those are diseases."

Anderson's caution is based on experience. In 1990, he and his colleagues at the National Institutes of Health in Bethesda treated a four-year-old girl suffering from ADA deficiency, a condition that creates severe problems in the immune system, with the then novel approach of gene therapy. Blood was drawn from the girl, genes responsible for regulating ADA production were inserted into the blood cells using a retrovirus as a delivery vehicle, and the blood was injected back into the little girl. The therapy was successful, but the researchers soon discovered that the process had to be repeated periodically to keep the girl's immune system functioning properly. A decade later, Anderson is the first to admit that gene therapy remains a great laboratory experiment not easily translated into reliable reproducible medicine, a fact that was driven home by the first death of a gene therapy patient, eighteen-year-old Jesse Gelsinger, on September 17, 1999. "The unfortunate fact is that, with the exception of a few anecdotal cases, there is no evidence of a gene therapy protocol that helps in any disease situation. Our bodies have spent tens of thousands of years learning how to protect themselves from having exogenous DNA get into their genomes. So we were a little naive to think that if we made a viral vector and put it into the human body it would work. The body's done a very good job of recognizing viral sequences and inactivating them," Anderson said in 1998.

Still, like most other researchers, he speaks as though he believes that germline gene therapy will be much easier to implement than somatic gene therapy, the method that's been tried so far. What's the difference? Somatic gene therapy involves the inserting of genes into the body of a child or an adult in an attempt to cure or ameliorate a disease. Germline engineering or gene therapy involves manipulation of egg and sperm (germ cells) and the fertilized egg, with scientists inactivating or replacing a disease-inducing gene. When genes are deleted, altered, or enhanced in the fertilized egg, those changes will be copied into every cell of the growing organism. Moreover, the change will be passed on to future generations because the primary material of heredity, the germ cell, has been altered. Somatic cell gene therapy, by contrast, is not heritable. And it is extremely difficult, a real hit-or-miss proposition, to insert genes into a fully formed organism. Some "take," some don't. The technical aspects of germline engineering are actually much simpler and much easier to implement. Yet, simple as the process might appear, the fact remains that no one can predict the outcome. "Our genes already know how to work in the context of many, many other genes," said Mario Capecchi, creator of the first "knockout" mice, in which certain genes have been deactivated, at the "Engineering the Human Germline" symposium. "One thing we must appreciate is that interactions of genes are going to change all the existing information. We have to become wise enough to know how the interaction is appearing."

Scientists are learning quite a lot about certain diseases caused by a mutation in a single gene, and these will doubtless someday be relatively "easy" to alter. But do researchers know enough to predict how the many genes involved in a complex "trait" like intelligence, or in diseases governed by multiple genes, will behave when edited? "Most traits of medical relevance do not follow simple Mendelian monogenic inheritance. Such 'complex' traits include susceptibilities to heart disease, hypertension, diabetes, cancer, and infection," commented the authors of a 1994 review titled "Genetic Dissection of Complex Traits." "To some extent, the category of complex traits is all-inclusive. Even the simplest genetic disease is complex, when looked at closely." Nonetheless, as the authors admit, "the genetic dissection of complex traits is attracting many investigators with the promise of shedding light on old problems and is spawning a variety of analytical methods."

Genetics and genetic medicine are hot fields right now, luring both researchers and the public with their technical virtuosity and pizzazz. The science itself is very sexy, guaranteeing years of work on undeniably fascinating problems. But the difficulties and dangers are also real, and easy to overlook in the current climate, when most people seem once again willing to leave science to the scientists, at least in the United States. Information is being produced at an incredible volume, long before we have any idea what to do with it. At a 1999 lecture titled "Ethics and Genetics: From Testing Ourselves to Designing Our Children," bioethicist Erik Parens of the Hastings Center discussed the difficult questions raised by the kind of prenatal testing that has become ubiquitous in the United States and other developed nations. Today prenatal tests are done in the second trimester of pregnancy, but soon maternal-fetal cell sorting will be available at eight to nine weeks. This may well be a good thing, since the new method will be quicker and cheaper and will reduce the risk of miscarriage associated with current testing. Nonetheless, "it signals a move to more and more testing," Parens said, "at a time when people are beginning to express strong reservations" about the implications of current practices. Some of these issues have already been discussed at length by medical sociologist Charles L. Bosk in *All God's Mistakes: Genetic Counseling in a Pediatric Hospital.* Bosk spent years observing the work of genetic counselors, who must explain the results of genetic testing to parents who are then faced with extremely painful decisions about abortion, postbirth surgeries, and ameliorative care, child rearing, and future childbearing.

"I was overwhelmed by the enormity of the kinds of decisions that parents had to make," Bosk wrote. "I was overwhelmed as well by the neonatal intensive care unit. I had trouble coming face to face with the various forms that genetic defects take, observing physical exams was always a personal trial. In addition, I was troubled by a sense of the incommensurable. What I observed, what I recorded was so prosaic; yet the implications of advances in genetic science, which promises to revolutionize medical practice, are anything but ordinary."

Admiring "the breathtaking technical competence" that characterizes clinical genetic services, Bosk nevertheless found himself "disturbed by the fact that the everyday practice of genetics was so unobjectionable in the early 1980's and that it remains largely unobjectionable today. As Troy Duster argues, the scientific practice of genetics by physicians in the community keeps open a *Backdoor to Eugenics*. Recent developments only pry the back door open wider. . . . The fact that this back door is ajar worries me, as does the hubris of the enterprise."

Bosk connected the mundane nature of the work being carried out every day in hospitals, the series of tests we now expect pregnant women to undergo, with the metaphysical implications of the need to control that such ordinary activities indicate. "The mapping of the human genome, the prenatal screening of fetuses for genetic fitness, the therapeutic manipulation of our genetic make-up—all of these are audacious exercises, even if they are also a part of a Kuhnian normal science. I cannot tell which is the greater hubris, that we try to do these things, that we do them with so few second thoughts, or that we do them despite the magnitude of second thoughts," he said.

The disability community in particular, bioethicist Erik Parens commented, has raised three important objections to the practice of prenatal testing, followed by abortion of fetuses deemed potentially defective or failure to carry out surgeries and other treatment that would extend the lives of seriously disabled infants. "First, there's the misinformation argument," Parens said. "That people fear having a child with a disability because they don't understand what life with a disability is really like. Second, there's the expressivist argument—that testing itself sends an ugly message to and about people with disabilities. And finally, there's a parental attitude argument, that the proliferation of testing is promoting loony ideas about how much control we should have over our children. There is a concern about where we as a society are going with this technology."

Some observers foresaw the difficult questions that these technologies would raise regarding human life and human nature very early on, and recognized their similarity to quandaries that humans have faced in the past, when confronted by the potential consequences of human ingenuity. "Medical advances are not just scientific and technological developments but also moral events, because they present humans with inescapable decisions about what we ought to know and do. New knowledge extends power, and new duties accompany these powers," theologian Robert Esbjornson wrote in the introduction to the published proceedings of a Nobel conference, "The Manipulation of Life," held in October 1983. "New powers are complex moral burdens. They increase control at the same time they increase the risks of losing control."

IN A STUDY of the history of human experimentation in America before the Second World War, titled *Subjected to Science*, medical historian Susan Lederer observed that "developments in the medical sciences created new opportunities and demands for human experimentation" at the beginning of the twentieth century. She reveals that animal protectionists were the first to observe that experimentation on animals was related to experimentation on humans. "Already devoted to saving dogs, cats, and other animals from the vivisector's knife, antivivisectionists warned that the replacement of the family physician by the 'scientist at the bedside' would inspire nontherapeutic experimentation on vulnerable human beings." Nineteenth- and early-twentieth-century antivivisectionists believed that certain groups were particularly vulnerable to being used as research subjects— children (particularly orphans), the poor, inmates, minorities, and military personnel. As Lederer documents, these fears, which the antivisectionists shared with the general populace, were amply borne out by scientific practice, even before the Nazi medical atrocities of World War II.

The prosecution of German physicians and scientists by an American tribunal after the war led to the promulgation of the Nuremberg Code to protect the rights of human subjects of biomedical experimentation. This is the code's first principle: "The voluntary consent of the human subject is absolutely essential. This means that the person involved should have legal capacity to give consent, should be so situated as to be able to exercise free power of choice, without the intervention of any element of force, fraud, deceit, duress, overreaching, or other ulterior form of constraint or coercion; and should have sufficient knowledge and comprehension of the elements of the subject matter involved as to enable him to make an understanding and enlightened decision."

Bioethicist George J. Annas has described the attempts of the international research community to dilute this strong statement of the primacy of informed consent in all biomedical research in the years after the war. "Physician-researchers viewed the Nuremberg Code as confining and inapplicable to their practices, because it was promulgated as a human rights document by judges at a criminal trial, and because the judges made no attempt to deal with clinical research on children, patients, or mentally impaired people. The World Medical Association has consistently tried to marginalize the code by devising the Declaration of Helsinki, a more permissive alternative document, first promulgated in 1964 and amended three times since," Annas said. "The Declaration's goal is to replace the human rights–based agenda of the Nuremberg Code with a more lenient medical ethics model that permits paternalism."

As Annas points out, the Declaration of Helsinki makes a distinction between therapeutic and nontherapeutic research and eliminates the need for consent in certain types of therapeutic research. The Declaration of Helsinki also replaces the Nuremberg Code's unequivocal assertion of the necessity for informed consent with peer review of research protocols. "The movement to displace the consent requirement of the Nuremberg Code with prior peer review by a medical committee found its greatest success in the 1975 Declaration. For example, the physician need not obtain the subject's informed consent to medical research combined with professional care if he submits his reasons for not obtaining consent to the independent review committee."

Informed consent is a little-discussed concept when it comes to genetic engineering. For example, during a discussion at the Second Annual Congress on Mammalian Cloning, the panelists were asked by the author to address the issue of informed consent with respect to cloning and germline genetic engineering. Molecular biologist Lee Silver, a proponent of a market-driven approach to the new technologies, who compares genetic manipulation to boost intelligence to purchasing a private school education for one's child, replied, "Did you consent to be born?" Attendees at the "Engineering the Human Germline" symposium held in Los Angeles a few months earlier were a bit more sensitive to the question, although even at that conference a majority of the panel members discounted the need for informed consent. Anticipating Lee Silver's reply at the cloning congress, Daniel Koshland, Jr., a former editor of *Science*, said, "I'm not sure informed consent is always necessary. When I was a kid I didn't really have an option whether I should go to school or not. My parents told me to go. And I told my children. My children didn't have to a vote on who their mother was when we decided to have children. So I think, sometimes, to extend informed consent to the embryo is really a sort of theoretical construct."

These comments reveal the distinction between the more rigorous Nuremberg standard of informed consent, which forbids experimentation on subjects unable to convey assent, and the "paternalistic" Helsinki version, in which scientists and physicians operating under the constraints of peer review are permitted to make decisions for patients and research subjects. Even more problematic is attempting to define the subjects whose consent must be given. Broadly speaking, human society itself will be the "subject" of the kind of experimentation envisaged by advocates of market-driven genetic engineering—and the public has barely begun to understand the possibilities, much less offer informed consent to the kinds of research currently underway. Intrascientific conferences like the one held in Washington, D.C., in 1998 are usually not heavily covered in the popular press and do not generate much discussion outside the community of scientists. Then, too, times have changed, and the kind of community activism displayed in the wake of the recombinant DNA controversy and the animal rights protests of the late seventies has given way to an ironic acceptance of the double-faced nature of scientific power. On the one hand, biomedical science and technology confer life, health, and comfort, and on the other, death, penury (for the uninsured), and excruciating decisions about where life begins and ends and when it is worth preserving. Few seem willing to assume the painful task of sorting through these issues today and balancing gain with loss.

Nonetheless, the issue of informed consent looms large when one considers the individuals who will be the subjects of the next round of human experimentation—those children, yet unborn, who will undergo the kind of genetic manipulation proposed by people like Lee Silver, and the germline engineering proposed by less radical physicians. Even those who are willing to grant the benefits of genetic engineering to eliminate disease pause when contemplating the possibility of parents "enhancing" the genome of their offspring, attempting to program certain traits like intelligence or good looks into their children, or cloning children to replace beloved family members. It sounds like a bad science-fiction novel, but as the discussions at the 1998 cloning congress reveal, such scenarios are recognized as potential applications of the development of cloning technologies.

Like animals, unborn children have no voice, and their consent to such procedures is assumed. However, unlike animals, these children will one day speak—and what, one wonders, will they say? Some philosophers, ethicists, and lawyers have already raised the specter of "wrongful life" claims. The notion of wrongful life posits that it is immoral to bring a fatally defective life into being. If Mary Shelley were writing today, she might well have her monster pursue such a legal claim against his creator. But even assuming that the kinks in the technology are worked out on animals before their implementation in human beings, and that perfect children are produced, what might be the effects of allowing parents to "build" a child the way they build a house, choosing preferred traits the way they do kitchen cabinets or rug colors?

"Nobody wants to have an average child, of course," Lee Silver said at the symposium on human germline engineering, explaining why parents might want to engineer certain abilities and traits in their offspring. But what exactly is "an average child" if every child is a unique individual with his own blend of traits, abilities, talents? And won't the desire to program specific traits into one's child exacerbate the already destructive tendency of many narcissistic adults to force their children to conform to preconceived standards of excellence and perfection? If a child knows that his or her parent has invested enormous sums of money in making him or her somehow "better" than other children, won't that create a great deal of anxiety in the child to live up to his parents' grandiose expectations?

Psychologist J. Keith Miller has pointed out that "each child is by nature valuable, each child is by nature vulnerable, each child is by nature imperfect, each child is by nature needing, wanting, and dependent, and each child is by nature immature. . . . When parents don't respect and nurture these traits properly, and instead try to force the child to meet the parents' expectations, then the child's emotional and spiritual development is arrested. . . . Even though the parents may be well-meaning and the expectations not conscious, their dominating attempts to mold the child into being some preconceived 'type' are abusive. When a child is not assisted in developing mature mechanisms in the areas of its five natural characteristics, and is instead constantly 'shaped' toward the parents' unnatural (for the child) expectations, child abuse is said to have taken place."

From this definition, it seems that some types of genetic manipulation being contemplated today might indeed be considered child abuse, if implemented. Giving individuals obsessed with shaping, rather than nurturing, their children yet another means to do so seems just as shortsighted as giving societies that have increasingly developed coercive definitions of "normalcy" another means to control their populations. Genetic "normalization" now threatens to succeed behavioral and psychological normalization. But who decides what's "normal," and where did the concept of "normal" come from? When did people begin to think about what constituted an "average child"—or an average adult, for that matter? And why, at the close of the twentieth century, do we devote so much of our prosperity and purpose to the quest to overcome the burdens of an "average" human existence, including an ever more feverish search for perfect health, appearance, and psychological functioning?

As the writer Daniel Callahan has noted, "Few causes or crusades have such universal support as medicine's war against suffering. None of us wants to be sick or to be in pain. Most people do not want to die. Yet we rarely ask when enough is enough in waging that war." Callahan notes that "biomedical research—nicely symbolized by the research agenda of the National Institutes of Health—has set its face against every medical evil: disease, disability, death. While many compassionate doctors are working hard to help people accept death as an inevitable part of life, the research juggernaut resists. It has targeted death and illness as horrors to be implacably fought." The enormous amounts of money spent on health care in the United States, "$1 trillion annually, 14% of our gross national product and far more per capita than any other country," according to Callahan's 1998 article, indicate that "no matter how much health status improves in this country—it is now at an all-time high—it is never good enough."

Arguing that "the crusade against any and all suffering can itself turn sour," Callahan proposes that "we would be much more fully served by a practice of medicine that knew its own limits in triumphing over suffering—limits to the benefits of research, limits to the struggle against death, limits to relieving our pain and misery. . . . If we could learn to say that enough is enough, we might open the way to a fresh exploration of human suffering. We might bring medicine back to a better balance between the desperate search for cures and the enduring human need for care and comfort."

But if science and medicine by their very nature reject such limits, who will set the boundaries? And why has the urge to transcend the seeming limits of our biology become such a mania? "Scratch the surface of both the information and biotech revolutions . . . and what one discovers underneath is a 'control revolution,'" science writer David Shenk suggested in an essay titled "Biocapitalism: What Price the Genetic Revolution?" published in *Harper's Magazine* in 1997. Both public and academic suspicion of science coalesce around this issue of control, manifested in the scientific domination of nature and the disciplining of human biology and behavior. A rejection of scientific authority and an embrace of other forms of knowing and experiencing the world unite many critics of reductionist rationality, despite their differing orientations, and form the crux of the postmodern critique of science.

TEN

ANIMALS, SCIENCE, AND THE BODY

Rationality, when conceived as complete, as excluding all arbitrariness, becomes itself a kind of irrationality.
—Arthur O. Lovejoy,
The Great Chain of Being, 1936

A BABY BOY IS BORN in a hospital near the close of the twentieth century. The infant is weighed and measured. Even before he is placed in his mother's arms, he has passed his first test—an assessment called the Apgar scale, which scores his response to stimuli and his motor functioning. A few hours later, his pediatrician visits for the first time, to have a look at his new patient and to perform a first examination. The next morning in the hospital nursery, the foreskin of the infant's penis is removed. That afternoon he goes home with his parents, locked into a car seat to protect him from sudden stops and the unreliable human embrace of his mother.

Over the next few months as he grows rapidly, his parents regularly consult the physician-authored manuals they have bought or received as gifts—Spock, Brazleton, and their peers. The books offer comfort and reassurance, helping the nervous new parents decipher the meaning of the infant's crying, confirm the regularity of his bowel movements, and establish the proper feeding schedule. The books tell the infants' parents at what age he should begin to hold his head up, eat solid food, roll over, crawl, walk. Every month the child is examined by the pediatrician during a well-baby visit, to ensure that he is developing properly. Whenever he starts to sniffle, cries more than usual, or in any way deviates from what his parents believe is "normal" behavior, they whisk him off to the pediatrician. The doctor, an older man near the end of his career, often reminds the mother how fortunate she is to have a child now, when most childhood illnesses are easily treated. This has not always been the case, though by now many have forgotten the fact.

When the boy enters preschool at four, people comment that he is big for his age. Within a few months, his parents worry that he seems to be more aggressive than the other children. On the semester- and year-end reports his preschool teacher writes and sends home, she notes that while his large motor skills are age-appropriate, his small motor skills are poorly developed. He has difficulty holding a pencil and forming numbers and letters. Although many boys his age have trouble sitting for long, this child is more restless than most, wriggling out of his chair long before activities are concluded. This pattern persists throughout preschool and kindergarten. By the time he begins first grade, he has been diagnosed as suffering from attention deficit disorder and is taking the drug Ritalin to help him calm down and focus on his schoolwork.

NINETY PERCENT of the more than twelve million kilograms of Ritalin produced in 1997 was consumed by American school-children. According to their physicians, these children suffer from attention deficit/hyperactivity disorder (ADHD) or attention deficit disorder (ADD), hazily defined neurological syndromes characterized mainly by unmanageability and an absence of self-restraint. Parents who struggle, sometimes for years, to raise such children are grateful for the relief the drug provides. But many critics question why there has been such a sharp rise in the number of children diagnosed ADD/ADHD, why most of those children fit a certain socioeconomic profile (white, upper-middle-class), and why most of them are American. Is the definition of ADD/ADHD and its consequent treatment with Ritalin and other drugs a benefit of affluence or a curse? Perhaps ADD/ADHD is the psychological equivalent of polio—a disease that makes its presence felt only in conditions of relative prosperity, with its attendant lifestyle changes. And if we were to follow the prototypical child described above throughout his life, what would we see? Is the script already written, based on this brief summary of his early years? If by first grade he has already been defined as disruptive, a bit of a troublemaker, defective in some key (albeit pharmacologically manageable) way, isn't the child's identity already established? In some crucial way, he is not "normal," although a drug can help him function in a world that demands at least the illusion of normalcy.

These troubling questions reflect the public face of a debate that has been taking place over the past thirty years in academic circles, much of it inspired by the seminal work of the French philosopher of history Michel Foucault, on the increasingly stringent and coercive definitions of normalcy (both physiological and psychological) developed in Western culture over the past three hundred years. In *The Birth of the Clinic, The History of Madness, Discipline and Punish, The History of Sexuality*, and other works, Foucault sought to unearth the intellectual artifacts that testify to the construction of our modern definitions of normal behavior and being. "Foucault seeks to show how the 'classical age' (more commonly termed the Enlightenment) was distinctive not for its faith in intellectual liberation but rather for its commitment to the disciplining of human behavior," comments the scholar Patrick H. Hutton. "Foucault's early authorship represents an inquiry into the policing function as it is understood in the French sense: the disciplining of human affairs by public and quasi-public agencies. Foucault's thesis about the policing progress is the key to his understanding of the psyche as an abstraction conjured up by public authority to satisfy the needs of modern society for a more disciplined conception of the self. It is through the policing process, Foucault contends, that the modern frame of mind has been formed."

Modern medicine and public health have played a significant role in this process of normalization, according to Foucault and his intellectual heirs. "Medicine must no longer be confined to a body of techniques for curing ills and of the knowledge that they require; it will also embrace a knowledge of *healthy man*, that is, a study of *non-sick man* and a definition of the *model man*," Foucault wrote in *The Birth of the Clinic: An Archaeology of Medical Perception*. "In the ordering of human existence it assumes a normative posture, which authorizes it not only to distribute advice as to a healthy life, but also to dictate the standards for physical and moral relations of the individual and of the society in which he lives."

Paradoxically, this trend toward normalization has proceeded in tandem with the growth of individualism, the emphasis encoded in post-Enlightenment political systems on the individuals as essential to the creation and maintenance of the modern state. The state's focus on individuals and on populations of individuals, together with the need to provide for their physical well-being and regulation, is called by Foucault "biopolitics," and the function of the "police" (a word Foucault uses not in the contemporary American sense, but in a much older, broader European sense) is to order and regulate all types of social relations.

"From the heart of the Middle Ages power traditionally exercised two great functions: that of war and peace, which it exercised through the hard-won monopoly of arms, and that of the arbitration of lawsuits and punishment of crimes," Foucault pointed out. "Now in the eighteenth century we find a further function emerging, that of the disposition of society as a milieu of physical well-being, health, and optimum longevity. The exercise of these three latter functions—order, enrichment, and health— is assured less through a single apparatus than by an ensemble of multiple regulations and institutions which in the eighteenth century take on the generic name of 'police.' The sudden importance assumed by medicine in the eighteenth century originates at the point of intersection of a new analytical economy of assistance with the emergence of a general 'police' of health."

Seeking to express the contradictions inherent in the creation of the modern person, Foucault found that the very process that makes the individual subject to a Power outside himself also helps him craft an individual Self; that somehow the Power that seeks to define and thereby constrain the individual actually creates wholly new possibilities for self-definition and resistance. "Power would be a fragile thing if its only function were to repress, if it worked only through the mode of censorship, exclusion, blockage and repression, in the manner of a great Superego, exercising itself in a negative way," Foucault wrote. "If, on the contrary, power is strong this is because, as we are beginning to realize, it produces effects at the level of desire—and also at the level of knowledge. Far from preventing knowledge, power produces it. If it has been possible to constitute a knowledge of the body, this has been by way of an ensemble of military and educational disciplines. It was on the basis of power over the body that a physiological organic knowledge of it became possible."

In the world according to Foucault, power makes us want the things it thinks we ought to want. Power makes us want to be the things it thinks we ought to be. If power says that being normal means being able to sit in a classroom for six hours a day at age six, then we (and our parents) will internalize that requirement and view anything other than the desire and the ability to sit still and work quietly and cooperatively as abnormal. If power thinks that a woman whose height is five feet four inches should weigh 120 pounds, then women of that height will struggle to achieve and maintain that weight. "Where power works 'from below,' prevailing forms of selfhood and subjectivity (gender among them) are maintained, not chiefly through physical restraint and coercion (although social relations may certainly contain such elements), but through individual self-surveillance and self-correction to norms," writes philospher Susan Bordo. "There is no need for arms, physical violence, material constraints. Just a gaze. An inspecting gaze, a gaze which each individual under its own weight will end by interiorizing to the point that he is his own overseer, each individual thus exercising this surveillance over, and against himself."

The wholly new subject that is both created through this process and repressed by it is a preeminently physical self, a body. Suddenly, strangely, after centuries in which the body was viewed as the soul's albatross, an unreliable and potentially dangerous container for the true essence of the individual, the body has become in modern and postmodern thought the seat of personhood, a primordial materiality constrained and repressed by a normalized and normalizing "soul." "The soul is the prison of the body," Foucault famously asserted. Much could be said about the genealogy of this claim, its debt to Freud, Nietzsche, and other thinkers. However, it is critical to understand this statement, and, more generally, the fetishization of the body in postmodern studies, as a product of science, particularly the biomedical sciences, which have over the past two hundred years increasingly defined the person *as* the body. This development is currently being taken to its apogee in contemporary genetics research, which claims the source of qualities such as intelligence, kindness (defined as altruism), aggression, and other states of being in the body—specifically, in genes. However, the progressive discounting and gradual diminishment of that numinous nonmaterial entity once called "soul," and a corresponding emphasis on that which could be seen, felt, and measured by science—material reality, the body—was set in motion in the eighteenth century, when rules of distant, detached, so-called objective observation practiced in the physical sciences began to be adopted in biology, particularly in medicine. Once scientists began to focus on human beings as matter, a complex yet entirely comprehensible interplay of organs, tissues, and cells and later, proteins and enzymes, nonscientists, too, adopted scientific perception and language and began to see themselves as purely material entities, as simply bodies—although this process remains incomplete, as we shall see.

Describing the complex relationship between subjects and the powers that mold and shape them, Judith Butler, a feminist philosopher influenced by Foucault, notes that "as a form of power, subjection is paradoxical. To be dominated by a power external to oneself is a familiar and agonizing form power takes. To find, however, that what 'one' is, one's very formation as a subject, is in some sense dependent upon that power is quite another. We are used to thinking of power as what presses on the subject from the outside, as what subordinates, sets underneath, and relegates to a lower order. This is surely a fair description of part of what power does. But if, following Foucault, we understand power as *forming* the subject as well, as providing the very condition of its existence and the trajectory of its desire, then power is not simply what we oppose but also, in a strong sense, what we depend on for our existence and what we harbor and preserve in the beings that we are."

Foucault's notions of power and its role in the formation of the subject are useful in considering modern conceptions of science, nature, and the body. As Susan Bordo has noted, "Since the seventeenth century, science has 'owned' the study of the body and its disorders. This proprietorship has required that the body's meanings be utterly transparent and accessible to the qualified specialist (aided by the appropriate methodology and technology) and utterly opaque to the patient herself. It has required, too, the exorcising of all pre-modern notions that the body might obey a spiritual, emotional or associational rather than a purely mechanical logic. . . . In the medical model, the body of the subject is the passive tablet on which disorder is inscribed."

The relevance of this analysis to the scientific and technologically defined world we children of the Enlightenment inhabit at the beginning of the twenty-first century seems clear. We are all, animal and human, subjects formed by science, the Power that orders and defines our lives and our world. Even those individuals seeking to escape the embrace of science, the most explicitly antiscience and technophobic deep ecologist, animal rights activist, or ecofeminist are subjects whose very resistance and rebellion is authored in and by the power that they abhor. As even Paul R. Gross and Norman Levitt admit in their polemic against postmodern critiques of science, *Higher Superstition: The Acade-*

mic Left and Its Quarrels with Science, science "has all of us by the throat."

In contemporary Western culture, it is nearly impossible to escape the normalizing presence of biomedical science, which now structures every aspect of human life and identity. From birth we are weighed, measured, and evaluated against "objective" criteria based on numerical averages and research data acquired by a number of scientific and social scientific disciplines. We are deluged by a steady stream of often conflicting research-anchored advice on the proper diet, the value of exercise, and the potential for disease lurking within the dark recesses of our bodies. The scientific-medical "gaze" named by Foucault reaches today into our very genes. "The clinician's gaze becomes the functional equivalent of fire in chemical combustion; it is through it that the essential purity of phenomena can emerge: it is the separating agent of truths," Foucault wrote. "The clinical gaze is a gaze that burns things to their furtherest truth."

An article by Malcolm Gladwell on the Ritalin controversy published in *The New Yorker* in 1999 amply illustrates this observation. Summarizing the objections of some authors to the prescription and possible overprescription of Ritalin and the related suspicion of the diagnosis of ADD or ADHD, Gladwell turns to science to settle the question. He describes video-game tests carried out by behavioral scientists on normal children and those diagnosed with ADHD. The tests show that the ADHD kids "completed fewer levels and had to re-start more games than their unaffected peers." Other tests show that ADHD kids take longer to read five rows of letters consisting of five levels of letters repeated in different order. This simple quantitative evidence is supported by clinical data. ADHD "turns out to have a considerable genetic component. As a result of numerous studies of twins conducted around the world over the past decade, scientists now estimate that ADHD is about seventy percent heritable," Gladwell says. Finally, and most impressively, he turns to molecular biology, which indicates that ADHD may be a simple matter of too little dopamine, a neurotransmitter that seems to "play a major role in things like attention and inhibition."

Postmodern critics of science infuriate scientists by questioning the entire intellectual apparatus that, in the manner described above, identifies and treats diseases like "attention deficit hyperactivity disorder." Insisting that science is produced by human beings, cultural constructivists seek the philosophical, economic, and social rationales for the creation of such a disease rather than accepting the disease itself as a "fact" discovered and explained by scientific disciplines like genetics and neuropharmacology. Postmodernists strike at the heart of the scientific enterprise, questioning the physical reality on which science is based, suggesting that the material world that science studies is an open-ended and infinitely plastic universe open to a variety of interpretations. More aggressively, postmodernists insist that science itself creates the phenomena that it purports to study. This radical skepticism strikes most scientists as a close cousin of insanity—that, or a canny attempt to overthrow the cultural supremacy of science. "The central ambition of the cultural constructivist program—to explain the deepest and most enduring insights of science as a corollary of social assumptions and ideological agenda—is futile and perverse. The chances are excellent, however, that one can account for the intellectual phenomenon of cultural construction *itself* in precisely such terms," say Gross and Levitt.

THE TRUTHS OF SCIENCE, like all arcane truths (including that of its academic adversary, postmodernism), are expressed in a special language. Since the nineteenth century that language has grown increasingly opaque, nearly indecipherable to those outside the specialized disciplines that today constitute the branches of scientific knowledge. Foucault discussed the development of the language of biology and medicine, a language that grew in tandem with the forms of knowledge it sought to express, in *The Birth of the Clinic*. "Over all these endeavors on the part of clinical thought to define its methods and scientific norms hovers the great myth of a pure Gaze that would be pure Language: a speaking eye. It would scan the entire hospital field, taking in and gathering together each of the singular events that occurred within it; and as it saw, as it saw ever more and more clearly, it would be turned into speech that states and teaches, the truth which events in their repetitions and convergence would outline under its gaze would, by this same gaze and in the same order, be reserved, in the form of teaching, to those who do not know and have not yet seen. This speaking eye would be the servant of things and the master of truth."

The "speaking eye" of science has become, over the past three hundred years, not only the dominant vision and language of our culture, but in the way of all totalizing systems, in its own eyes the sole legitimate vision and language. For scientists and those who subscribe to the scientific ethos, people who describe the world in terms other than those of the rational mechanistic scientific model are not just different but deluded. Science sees the world as it is, beyond argument, according to science. Other systems, other values, other languages may entertain or annoy, but they have no power to affect the basic truths of science, expressed in a language that discriminates between initiates and noninitiates. If one cannot speak the language of science, does not have access to the power and truth expressed in and through that language, then one is not really capable of "rational" communication at all. One version of this belief is presented by Gross and Levitt in *Higher Superstition*, in their argument that only scientists are capable of analyzing and critiquing the practice of science and its interactions with culture.

"A serious investigation of the interplay of cultural and social factors within the workings of scientific research in a given field is an enterprise that requires patience, subtlety, erudition, and a knowledge of human nature," the authors say. "Above all, however, it requires an intimate appreciation of the science in question, of its inner logic and of the store of data on which it relies, of its intellectual and experimental tools. In saying this, we are plainly aware that we are setting very high standards for the successful pursuit of such work. We are saying, in effect, that a scholar devoted to a project of this kind must be, inter alia, a scientist of professional competence, or nearly so."

Gross and Levitt, like many who object to outsider critiques of science, seem unaware that the very nature of the gifts that encourage a person to pursue a career in science, together with the specialized training that fine-tunes that innate competency and indoctrinates the prospective researcher into the culture of science, tends to eliminate (or at least suppress) other ways of responding to the world. Even the most sensitive graduate student learns very quickly to suppress his or her emotional response to the demands of animal research, for example. Deep empathy for other life-forms, while an asset for a poet or a painter, is a decided liability in a student of physiology. Moreover, prospective physicians and scientists are hardly encouraged to study history, philosophy, literature, and the arts. Even an undergraduate course of study in science is highly focused and does not offer much room for students to pursue those other forms of learning generally termed the liberal arts. This does not mean that science students are not free to pursue these interests in their spare time, but honestly, how many do so, or can afford to do so? Narrow specialization is amply rewarded in academia, while those whose work expresses an interdisciplinary perspective are often penalized, either subtly or overtly. And for the most part, only those who speak the language of science, which expresses a collectively affirmed and validated worldview, are trusted by fellow scientists.

"One now sees the visible only because one knows the language; things are offered to him who has penetrated the closed world of words; and if these words communicate with things, it is because they obey a rule that is intrinsic to their grammar," Foucault wrote in *The Birth of the Clinic*, describing the way that knowledge and control of the body, power over the body, is inextricably bound up with the language used to describe it. "Operational mastery over things is sought by accurate syntactic usage and a difficult semantic familiarity with language. Description, in clinical medicine, does not mean placing the hidden or the invisible within reach of those who have no direct access to them; what it means is to give speech to that which everyone sees without seeing—a speech that can be understood only by those initiated into true speech. . . . Here, at the level of theoretical structures, we encounter once again the theme of initiation. . . . We are at the heart of the clinical experience—a form of the manifestation of things in their truth, a form of initiation into the truth of things."

THE ROLE OF ANIMALS as subjects of scientific research, and the vision of what it is to be human that this methodology has created, once again leads us to an interesting paradox. Animals have been perceived as appropriate subjects of experimentation largely because they are not human, thus not protected by the same ethical constraints that forbid experimentation on human beings. Yet by using animals to construct a biomedically accessible (and definable) human subject, science has focused precisely on those aspects of human physiology and behavior that are most animal-like, ignoring that which does not conform to the model. For the past two hundred years, researchers have drawn parallels between the human and the animal. Initially, this work focused purely on the physiological, but increasingly over the past seventy-five years the psychological has been an equally intense focus of study.

With every new discovery, every experimentally confirmed similarity between an ape and a man, or a mouse and a man, scientific research has steadily eroded the once impenetrable barrier between animal and human. Bruce Alberts, president of the National Academy of Sciences, could barely contain his exultation when the full genetic code of the roundworm *Caenorhabditis elegans* was deciphered in 1998. "In the last ten years we have come to realize humans are more like worms than we ever imagined," he said in a front-page article in *The New York Times*.

Darwin's intuition of continuity has been proven many times over. At the same time, scientists have insisted that animal experiments are ethically permissable because animals are in some key way different (and lesser) than humans. Yet the methodology of animal experimentation (and to a greater extent, a worldview focused exclusively on the material) can never discover the nature of this crucial difference if it is something that transcends the material—if it cannot be localized in tissue, cell, or gene.

To phrase this another way, science's great success has been based on its exclusive and unwavering emphasis on the material. Yet that very success carries a related limitation—any extramaterial phenomenon, any event or process that cannot be observed and quantified using currently available tools and methodologies, cannot exist within the scientific paradigm. Moreover, unwilling to admit this limitation, science will work to convert all extramaterial phenomena into the material. Love, for example, will become a purely biological phenomenon, scientifically explicable on the basis of biochemistry or evolutionary imperatives or psychological dysfunctionality. Intelligence will become something that can be determined and quantified by tabulating a certain number of correct answers on a test. Spirituality will become something wholly other, completely unquantifiable and irrational and therefore inadmissible in a fundamentally scientific worldview. That animals do not appear to have spiritual strivings and that humans demonstrably do is not a question that an animal-based science of human physiology and psychology is prepared to confront, much less explore. And despite Darwin's pioneering work in *The Expression of the Emotions in Man and Animals,* the role and value of emotion in human life and decision making—and the parallels between human emotionality and the emotional lives of animals—remain largely unexplored except by popular writers whose work is viewed with distaste by most scientists.

Given these facts, resistance to the experimental use of animals in science (and to scientific instrumentalism in general) seems to grow from two separate, opposed sets of belief. One shares science's assumptions, and the other refutes them. The first accepts the scientific conflation of human and animal and uses it to construct a worldview based on the belief that humans are a kind of animal and should therefore inhabit the earth as one animal among many, with no special rights or privileges. This belief is generally expressed by those who espouse an animal rights philosophy or a radical environmentalism.

The second form of resistance emanates from the belief that humans and animals differ in some essential characteristic that is impervious to the scientific Gaze—that indeed much of what is most valuable and precious in human life is not amenable to scientific scrutiny. This belief is the locus of resistance in those who are opposed to science as a philosophy and who express their objections in literary or mystical terms.

Both perspectives can coexist in the same individual or group, in the same way that most scientists accept that animals are both like humans and other than human. Both perspectives also have long and distinguished histories in the West that run parallel to the development of the scientific method and the increasing cultural supremacy of the scientific worldview. Both can be read as attempts to resist the control of science, its quest to shape and define human beings and nature. The fierce desire to assert human freedom, over and above scientific determinism, underlies both the assertions of antivivisectionists that animal experimentation has produced nothing of value, and the more nuanced arguments of postmodern thinkers who attempt to prove that science is culturally constructed. Both are attempts to undercut the authority of science, an authority felt to be inimical to human freedom, and to resist a forced imposition of science's instrumental relationship to the natural world.

This desire to assert free will over scientific determinacy is consistently minimized and misunderstood by those seeking to defend the supremacy of the scientific worldview. What is generally condemned as "irrationality" by individuals committed to the type of reductionist rationality embodied in science may well be a completely rational response to the philosophical arrogance of science, an attempt to defend something precious and important from a worldview committed to the suppression of all that will not bow to its epistemology. The fact that those who resist the "totalizing" tendencies of science often construct their own totalizing systems of thought (for example, academic postmodernism, animal rights) presents an interesting paradox.

A GROUP OF WOMEN stands in a circle of trees at dusk. One holds a chalice, one a rock marked with a five-pointed star, another a wand made from the branch of a willow tree, and a fourth a candle whose flame flickers in the autumn wind. The full moon, a harvest moon, shines above the trees.

"I call upon the Guardians of the North, spirits of earth," says the woman with the rock, holding it in front of her at eye level and circling the group with the talisman held high.

"I call upon the Guardians of the East, spirits of air," says the woman with the wand, using it to trace an arc in the air as she follows the woman with the stone.

"I call upon the Guardians of the South, spirits of fire," says the woman with the candle, who joins the other two.

"I call upon the Guardians of the West, spirits of water," says the woman with the chalice filled with water, holding it high as she follows the other three in a dance that spirals in and out of the circle.

Candlelight flickers on the faces of the assembled group, who stand relaxed, yet serious and intent. They are of varying ages, from sixteen to sixty, dressed in street clothes but with unusual touches not usually seen on the street: feather earrings, necklaces in the shape of a birdlike woman, and, on a few, jewelry crafted in the same five-pointed star painted on the stone.

"We call upon the Guardians to join us in our ritual as we meet on this night of the full moon to honor the Goddess and to help heal the earth and her children," says the woman with the stone when she and the others have completed their circuit around the group.

"As we will it, so mote it be," answer the others.

"Goddess, come to us, show your face, and let us feel your love for us and for all your children. Show us the way to healing," the woman with the chalice says.

Then the group begins to chant, slowly at first, but then with increasing power, circling as they sing.

"Earth is my body, water is my blood, air my breath, and fire my spirit."

Over and over they repeat the chant, circling faster and faster until finally one of the women shouts "It is done!" and they pause with their arms lifted to the sky. As a flash of lightning illuminates the sky overhead and rain begins to fall, they drop to the ground, laughing.

Lambasting contemporary critics of science, Gross and Levitt, authors of *Higher Superstition: The Academic Left and Its Quarrels with Science* and editors of *The Flight from Science and Reason,* proceedings of a 1995 conference on antirationalism, target "the Goddess-worshiping camp" as one of science's enemies, without really defining what that camp might be or believe. Hurling most of the force of their argument against fellow academics working in the humanities, postmodern theorists influenced by figures such as Derrida, Foucault, Bordieu, and Lyotard, the authors of *Higher Superstition* focus on the adversaries they find most like themselves, therefore most threatening. "Goddess-worshipers," by contrast, are treated by the authors as too irrational, too marginal, to merit more than a passing mention that conflates neopaganism, a religion, with ecofeminism, a political philosophy. While there are certainly links between the two, not all neopagans are ecofeminists, nor all ecofeminists neopagans. Nor are all those who are seeking a new "feminized" spirituality— which reimagines the relationship between human beings, the divine, and the earth—adherents of nontraditional religions. Nonetheless, each of these approaches reiterates certain themes present in nineteenth-century antivivisection.

And now I show you a mystery and a new thing, which is part of
  the mystery of the fourth day of creation.
The word which shall come to save the world shall be uttered by
  a woman.
A woman shall conceive and bring forth the tidings of salvation.
For the reign of Adam is at its last hour; and God shall crown all
  things by the creation of Eve.
Hers is the fourth office; she revealeth that which the Lord hath
  manifested.
Hers is the light of the heavens, and the brightest of the planets
  of the holy seven.
She is the fourth dimension; the eyes which enlighten; the power
  which draweth inward to God.
And her kingdom cometh; the day of the exaltation of women.

Anna Kingsford's mystical holistic vision, evident in the above fragment from her *Prophecy of the Kingdom of the Soul, Mystically Called the Day of Woman*, has been reborn in the final decades of the twentieth century, in two movements that have attracted large numbers of women. The first, a political philosophy called ecofeminism, links the oppression of women to abuse of the natural world and attributes both to a Western tradition that devalues women's bodies and the bodies of animals, often conflating the two. The French writer Françoise d'Eaubonne coined the word *ecofeminisme* in her book *Le Feminisme ou la mort* (*Feminism or Death*), a title that neatly encapsulates the apocalyptic flavor of the movement. Karen J. Warren, a prominent ecofeminist theorist, defines ecofeminism as a philosophical position "based on the following claims: (i) there are important connections between the oppression of women and the oppression of nature; (ii) understanding the nature of these connections is necessary to any adequate understanding of the oppression of women and the oppression of nature; (iii) feminist theory and practice must include an ecological perspective; and (iv) solutions to ecological problems must include a feminist perspective."

Seeking to explain ecofeminism's rejection of scientific materialism and its claims to objectivity, Carol J. Adams, author of *The Sexual Politics of Meat* and other feminist analyses of animal rights, has written that "the same dominant mind-set that separates humans from the rest of nature divides politics from spirituality, as though humans are not part of nature and politics is not integrally related to spirituality." Susan Griffin, author of *Woman and Nature*, similarly asserts that "modern science, which in one way asserts the primacy of material knowledge, in a more subtle way preserves the same dividedness that preceded it. Using the scientific method, scientists attempt to be above sensual experience. But instead of being above experience, they are perceiving partially. They see the pieces clearly, with no feeling for the whole."

This theme (or elements of it) was often invoked by the women who made up the bulk of the membership of antivivisection societies in England and the United States in the nineteenth century. Contemporary women have used it to explore the links between their relationship to the dominant (male) culture and that culture's use and abuse of animals. "It could be argued that theorizing about animals is inevitable for feminism," note Carol J. Adams and Josephine Donovan, editors of a 1995 collection of essays on the subject. "Historically, the ideological justification for women's alleged inferiority has been made by appropriating them to animals: from Aristotle on, women's bodies have been seen to intrude upon their rationality. Since rationality has been construed by most Western theorists as the defining requirement for membership in the moral community, women—along with nonwhite men and animals—were long excluded.

Identifying three strains of feminist thought regarding the "historical alignment of women and animals," including one that disavows any attempt to identify women as closer to nature or more animal-like than men, Adams and Donovan argue that any feminist theory worthy of the name "must engage itself with the status and treatment of other animals." Rejecting liberal feminism, which attempts to provide women with rights and opportunities equal to those enjoyed by men in most societies, Adams and Donovan instead advocate "a broader feminism, a radical cultural feminism, which provides an analysis of oppression and offers a vision of liberation that extends well beyond the liberal equation, incorporating within it other lifeforms besides human beings."

These and other ecofeminist authors embrace an explicitly revolutionary approach. Characterizing feminism as "a transformative philosophy that embraces the amelioration of life on earth for all lifeforms, for all natural entities," Adams and Donovan link all forms of oppression into a single malignant force that threatens the survival of all. "We believe that all forms of oppression are interconnected: no one creature will be free until we are all free—from abuse, degradation, exploitation, pollution, and commercialization. Women and animals have shared these oppressions historically, and until the mentality of domination is ended in all its forms, these afflictions will continue."

The moral righteousness and all-encompassing compassion of this passage express an ethic that has traditionally been characterized as feminine and maternal and was explicitly articulated as such in the rhetoric of the nineteenth-century female leaders of the antivivisection movement. Despite their social radicalism, both Frances Powers Cobbe and Anna Kingsford accepted the Victorian view that women were especially well suited to campaign against the evils of vivisection by virtue of their inherent moral superiority. Unlike more traditional thinkers of the time, Cobbe and Kingsford believed that these superior moral qualities were needed outside the home in the broader realm of social affairs. Both Cobbe and Kingsford eventually withdrew from the overt struggle for women's rights to devote their energies to antivivisection. Their rationale was that women working to eradicate suffering and "to make society more pure, more free from vice, either masculine or feminine, than it has ever been before" would eventually reap far greater benefits for individual women and the sex in general than a single-minded focus on achieving political, economic, and social equity with men.

Clearly, this point of view continues to resonate with many women, who profess allegiance not only to their own personal advancement but to a more general reform of society, from "patriarchy" to "gylany," as feminist writer Riane Eisler has termed such a transformation, where the "dominator" values of a male elite are replaced by a female ethic of care, compassion, and relatedness. Even among relatively young women not identified with an explicitly feminist viewpoint, a greater empathy toward animals appears to be the norm. "Numerous studies have established that the gender of a person is the most powerful predictor of his or her general attitude toward animals," says Kenneth J. Shapiro, a psychologist writing about attitudes toward animal rights among students in *The Encyclopedia of Animal Rights and Animal Welfare*. "For example, one investigator found that in 10 to 15 countries studied . . . women significantly more than men opposed animal research. The reasons for this 'gender gap' are not fully understood but may involve differences in parental views of girls and boys, such as the importance given in the socialization of girls to developing caring and nurturing relationships." Whether this greater empathy and identification with others is biologically based or culturally constructed, the long-standing belief that women are more nurturant and less aggressive than men has been used both to praise women and to damn them.

This "essentialist" philosophy of gender relations—women are naturally more compassionate, men more aggressive—is not embraced by every contemporary author writing on these topics, however. Some ecofeminist writers explicitly reject this formulation and take a more balanced approach. In a 1992 book, *Gaia and God: An Ecofeminist Theology of Earth Healing*, Rosemary Radford Reuther, professor of theology at Garrett-Evangelical Theological Seminary, ascribes contemporary spiritual and environmental upheavals and cultural disenchantment with science and the scientific worldview to a number of factors other than gender. Reuther focuses on what she terms "the duel between the Christian biblical view of the world and that of the physical and biological sciences."

Reuther says that the tension between these opposing viewpoints had in many ways lessened over the past hundred years as each tacitly agreed to cede to the other separate spheres of authority. Scientists focused only on what could be empirically observed and measured, Reuther pointed out, and increasingly came to regard "all subjective matters of inward experience and value judgement, the ethical, aesthetic, and the spiritual," as not just unmeasurable but unreal. Religion, and the humanities in general, "could then establish its own separate reality in the subjective sphere."

But this agreeable compromise has begun to break down in recent years, Reuther argues, in part because the "triumphalistic assumption of science that religion was an obsolete superstition destined to fade away as science established its sure results, became tempered." The nuclear bomb and growing environmental problems, Reuther says, led people to question their trust in science and technology. By the final decade of the twentieth century, many had begun to believe that "the tools created by science might result in the destruction of the earth, rather than its decisive establishment on the road to prosperity and happiness for all."

Like many contemporary authors, Reuther also notes the inadequacy of science and scientific reductionism to satisfy the ancient yearnings of the human spirit "for aesthetic, ethical and spiritual dimensions of reality. These were implicitly, if not explicitly, rejected as unreal by the strict, scientific worldview," she says. New discoveries by scientists, particularly those of the quantum physicists as popularized by New Age writers, also began to erode the comfortable boundary between "objective" facts and "subjective" interpretation, calling into question the entire intellectual apparatus that supported the radical division between science and religion and the humanities. By the final decades of the century, Reuther argues, it was no longer quite "so clear that science could demarcate an objective realm of 'facts' distinct from subjective perspectives."

At present, Reuther notes, there is a pressing need for a new dialogue between science and religion to address the spiritual, environmental, and ethical crises we face in industrial-technological societies. "Culturally, this dialogue and synthesis have barely begun to gain currency," she says. Reuther's strategy for initiating this dialogue is quite conservative, as exemplified in her approach in *Gaia and God*, in which she evaluates the Judeo-Christian cultural tradition and its beneficial and destructive legacies. Unlike many ecofeminist authors, she does not valorize non-Western cultures, nor characterize our own as exclusively "patriarchal, hierarchal, and militaristic." Instead, she looks for sources of "ecological spirituality" within the Western Christian tradition and finds much to build on.

This perspective is similarly expressed by Matthew Fox, founding director of the Institute of Culture and Creation Spirituality at Holy Names College in Oakland, California. Fox, a former Dominican friar turned Episcopal priest after his silencing by the Catholic Church, is the author of fifteen books that (like Reuther's work) seek to anchor contemporary environmental and spiritual concerns within the Judeo-Christian tradition. In *Original Blessing*, published in 1983, Fox immediately poses two questions: In our quest for wisdom and survival, does the human race require a new religious paradigm, and does the creation-centered spiritual tradition offer such a paradigm?

Answering yes to both questions, Fox goes on to say that "there are two places to find wisdom: in nature and religious traditions. To seek wisdom in nature we should obviously go to those who have loved nature enough to study it. Because science explores nature it can be a powerful source of wisdom." Nonetheless, Fox notes that in Western culture, science and religion have been rivals, rather than allies, for three centuries. "This split has been disastrous for the people," he says. To heal this schism, Fox argues for what he calls a deep ecumenism, which moves beyond simple dialogue to creativity. "Ecumenism is not about talking together or putting out papers together," he says, "but about creating together. What can two parties, Protestant or Catholic, Christian or Buddhist, scientist or theologian, artist or mathematician, create together? That is the question that the universe and the human race, and God the Creator, put to all of us."

A similarly ecumenical stance is sometimes taken by adherents of nontraditional religions whose beliefs might seem totally at odds with the scientific worldview. For example, in her study of the growth of contemporary neopagan religions, *Drawing Down the Moon: Witches, Druids, Goddess-Worshippers, and Other Pagans in America Today*, National Public Radio journalist Margot Adler quoted surveys completed by subscribers to neopagan journals to illustrate the surprising fact that a majority of the journals' readers are employed in science and technology professions. Follow-up interviews confirmed these results, revealing that far from opposing science, most of Adler's sources supported science and technology—with a caveat.

"Turning to the question of 'scientific' vs. 'magical' thinking, almost everyone I questioned felt that there was no conflict between the two," Adler wrote. However, she noted that two key pagan theorists she interviewed were careful to make a distinction between scientific research and what they termed "the religion of science." Describing the latter as a form of fundamentalism with two primary dogmas—that it is not a religion at all, and that all other religions are superstitions—Adler's sources cited the dangers of this creed. "Unfortunately, the rebellion against the spiritual poverty of scientolatry often ends up becoming a mistrust of science itself and a mistrust of any kind of technological development. This is a confusion," said Aidan Kelly, a pagan priest who serves as one of Adler's primary sources throughout her book. "Science is a method, a technique. And technology is very useful. The problem is not with the tools, but with this idolatrous attitude toward science, this secular religion that denies all other aspects of being human."

The ability of the individuals interviewed by Adler to balance what appear to be absolutely incompatible perspectives—an appreciation for science and belief in an animistic universe—points to a less dramatic dichotomy that is familiar to most people. Human beings are not exclusively rational beings; each of us struggles to balance reason with emotion as we go about our daily lives. Most of us recognize that a life ruled solely by reason would be cold and barren, while one driven exclusively by emotion would be chaotic and unmanageable. Intuitively, we recognize the complexity of our nature, the human need and tendency to perceive and respond to the universe in a multitextured way. We are not driven purely by emotions nor by reason nor by instinct, but by some imprecise combination of the three. It is precisely this rather commonplace understanding of human nature that is ignored by the practitioners of what Aidan Kelly and others have termed scientolatry, the worship of pure reason.

Radical resistance to this worldview is not a new phenomenon, particularly among artists, writers, and other individuals committed both personally and professionally to what might best be termed culture—the customs, arts, and other characteristic products of human civilizations. As Gross and Levitt note, "It is in literature and poetry that we first begin to encounter a reaction against Enlightenment values that reveals a specific distrust of science, as well as a strong reluctance to believe that mankind can be reformed along 'scientific' lines." Mistrust of science (and scientific schemes to perfect humankind) in the nineteenth century among the Romantic poets in England and proponents of *Naturphilosophie* in Germany, two groups often assumed to be the originators of the antiscience aesthetic, does not necessarily imply a rejection of "Enlightenment values," however. How could that be, when the great proponents of Romantic individualism and idealism, such as Blake and Wordsworth, Schlegel, Schiller, Goethe, and Schelling, are as much a product of the Enlightenment as the scientists Newton, Pasteur, Bernard, and Koch?

Perhaps the rejection of scientific reductionism and instrumentalism expressed by the Romantics, the *Naturphilosophen*, and their modern-day heirs is just as much an expression of Enlightenment values (specifically a liberal humanism that continually resists efforts to circumscribe human freedom) as science itself. Perhaps the Romantic poets and philosophers who championed the power of individual vision and the possibility of intuitive communion with nature recognized that science, which proclaimed itself from the start as the great liberator (from orthodoxy, from superstition), had the potential to become the new arbiter of truth, the locus of a secular orthodoxy no less oppressive than the religious authority it sought to overthrow.

Another explanation for the long-standing refusal of artists and scholars in the humanities to succumb to an uncritical acceptance of science's grandiose self-image is that it appears self-evident to those who produce and/or study the artifacts of culture—literature, music, philosophy, art, and religion—that these are just as beautiful and essential a manifestation of the human mind, soul, intellect (whatever one chooses to call the indefinable something that makes us human) as science itself. Taking a historical view, modern science is a latecomer to the family of human cultural productions. Although it is clear that the desire to understand the natural world is a basic human aspiration, science as we know it was born less than four hundred years ago. It is worth remembering that the Inquisition burned early scientists as well as witches—both challenged the authority of the church by attempting to understand the natural world and to share their knowledge with others. Even at that early date, it was clear that such knowledge conferred power.

Notwithstanding its elegance and efficiency in describing and manipulating the natural world, amply illustrated over the past few hundred years, science alone probably cannot express the complexity and richness of the human response to nature and to other animals. The repeated assertion by fundamentalist advocates of philosophical materialism that science is uniquely insightful, uniquely valuable, uniquely capable of understanding and thus regulating human life understandably annoys those who give their allegiance to other gods. Moreover, it is perceived by many as a dangerous arrogance, one that threatens our survival, if it is not tempered by humility and an acknowledgment of its limits.

IN THE SAME WAY that power exerts itself on individuals from many directions at once, resistance to power is also multisourced and polyvocal. Over the past two hundred years, resistance to the growing power and normalizing tendencies of biomedicine has been exerted from a number of different sources, and groups have sometimes formed alliances against their perceived common enemy despite their inherent incompatability in aims and philosophy. But the pockets of resistance among certain interest groups has been supplemented by a growing estrangement between science and the humanities in universities, and between science and the educated public. W. M. S. Russell, the Oxford-trained zoologist who coauthored *The Principles of Humane Experimental Technique* in 1959, is a scholar in the humanities as well as a scientist and has written extensively on myth, literature, and social biology. In 1980, he published a three-part study of the link between biology and literature in Great Britain from 1500 to 1900, showing that "from the Renaissance to the mid-nineteenth century, biology and literature were closely related in Britain, and the culture of the educated reading public embraced both in an harmonious way."

However, by the latter years of the nineteenth century, Russell found "signs that large sections of the educated public were already becoming estranged from biology, and ceasing to find it familiar and comprehensible, at the time when education underwent its greatest expansion, so that biology was both neglected, and dissociated from literary studies, in the much enlarged school systems of the twentieth century." Russell identifies a number of culprits in this process, including the increasing opacity of scientific discourse. "At just the time when the educated public were beginning to lose touch with biology, some biologists, along with other scientists, were beginning to write for each other, and not for the public at all," he remarks, quoting a fellow scholar who noted that "this deplorable pseudo-English seemed to many people a new private code deliberately designed to keep them out."

While noting that these developments widened the gulf between biology and literature, Russell places responsibility for the schism mostly on public misunderstanding of evolution, and discomfort with animal experimentation. "While evolution caused vast intellectual difficulties, vivisection aroused strong emotions," Russell says. Adding that "sadistic and indiscriminate experimentation (which is generally scientifically worthless) has naturally revolted decent biologists as much as other decent people," Russell says that "an indiscriminate antivivisection message" in popular novels of the period contributed to the growing alienation of the public from biological science. Finally, he believes that prudery in the late Victorian and Edwardian eras also contributed to the reaction against biology and biological education, fed by a widespread squeamishness regarding discussion of reproductive biology and anatomy. Russell concludes his survey by noting hopeful signs of an increasing public interest in biology in our own times, and a burgeoning interest on the part of scientists in reaching out to the public. "By such means . . . I hope we shall see an end of the estrangement between biology and literature, and the revival of a wide public education in the sciences and arts, such as I have shown to prevail for nearly four centuries of British culture."

Russell's optimism about the possibility of a rapprochement between the arts and sciences and his success in establishing a philosophical position on vivisection acceptable to many animal protectionists and scientists (the three R's) are no doubt related to his own fluency in both science and the humanities. Like the science-affirming pagans quoted earlier, he seems able to recognize the value of both scientific epistemology and practice and its polar opposite, the valorization of subjectivity and emotional response embodied in the arts, myth, and religion—and to see both as valid and necessary components of human life. Russell contends that both routes lead to the same destination with respect to animal experimentation: "In my book with R. L. Burch on humane experimental technique, a virtually total positive correlation was established between the humaneness and the scientific value of experiments."

Partially because of the course of events set in motion by this discovery, and its eventual acceptance and implementation by policymakers in Europe and the United States, the war between antivivisection and animal research has given way to a tenuous truce—at least between those parties willing to accept the three R's solution proposed by Russell and Burch. However, although antivivisection itself no longer appears to be a prominent public issue (at least in the United States), it remains one aspect of a larger and more diffuse nexus of concerns about our science- and technology-based society, its impact on the well-being of the earth and its inhabitants, and our future course. Since the 1970s, the broad category of concerns labeled "environmental" have achieved ever greater prominence. Initially confined to concerns about clean air and water and the role of industry in pollution, environmentalism today in its broad definition implies a multi-disciplinary analysis and evaluation of the legacy of Western culture, the still resonating impact of scientific and industrial re-volutions—in effect, a critique of our entire way of life. Rational or not, a substantial number of people in the developed world appear to believe that the earth is in danger and that science is to blame. Another, smaller, group yearns for a return to a more "nat-ural" way of life, however imprecise the yearning and dangerously antidemocratic the desire to enforce that return may be. How does this current strain of environmental apocalypticism differ from past expressions of the Romantic aesthetic, acknowledged as a continuing manifestation of science-related anxiety?

First, this anxiety is no longer confined to an elite subset of writers and intellectuals who may be threatened by the increas-ing cultural authority of science, as in the past. A broad public consensus has been evolving over the past twenty-five years in the developed nations that the environmental crisis is real and must be addressed. As Luc Ferry, professor of philosophy at the Sorbonne, says in *The New Ecological Order*, a critique of radical environmentalism and animal rights, "the love of nature strikes me as essentially composed of democratic passions shared by the immense majority of individuals who wish to avoid a degradation in their quality of life; but these passions are constantly being claimed by the two extremist tendencies—neoconservative or neoprogressive—of deep ecology." Ferry notes that "the renaissance of feelings of compassion for natural beings is always accompanied by a critique of modernity—designated, depending on the frame of reference, as 'capitalist,' 'Western,' 'technological,' or more generally 'consumerist' "—and adds that "the critiques of modernity emanating from ecology movements are apt to be in opposition," that "deep ecology can combine in one movement both the traditional themes of the extreme Right and the futurist ideas of the extreme Left. The essential element which gives coherence to the whole is the heart of the diagnosis: anthropocentric modernity is a total disaster."

Ferry aptly characterizes the extremism of such a worldview by noting that its advocates call not for reformation but for a revolution. "The following are denounced, in order of their appearance in history: the 'Judeo-Christian tradition,' because it places the spirit and its law above nature, and Platonic dualism, for the same reason; the technical concept of science that triumphed in Europe beginning in the seventeenth century with Bacon and Descartes, for it reduces the universe to a warehouse of objects to serve man; and the entire modern industrial world, which gives priority to the economy over all other considerations. One cannot change the system by merely refitting it, as reformists naively believe: What is needed is a true revolution, including on an economic level, which implies that the critique of the modern world draws sustenance from radical principles."

But even those who reject revolution accept the need for change. Reams of scientific data have confirmed activists' intuition that the earth is suffering critical stress. Most prominent among the data-driven concerns are threats to biodiversity and significant climate change caused by human activity, potential shortages of water, loss of soil productivity, and pollution of the oceans. A significant number of scientists appear to share (and in some cases to promote) public anxiety about the environmental costs of current lifestyles and practices. A 1993 document with the apocalyptic title "World Scientists' Warning to Humanity," produced on flyers and over the Internet, has been signed by more than 1,670 scientists, including 104 Nobel laureates, many of them well-known and respected international figures.

"We the undersigned, senior members of the world's scientific community, hereby warn all humanity of what lies ahead. A great change in our stewardship of the earth and the life on it is required, if vast human misery is to be avoided and our global home on this planet is not to be irretrievably mutilated," the scientists maintain. "A new ethic is required—a new attitude toward discharging our responsibility for caring for ourselves and for the earth. We must recognize the earth's limited capacity to provide for us. We must recognize its fragility. We must no longer allow it to be ravaged. This ethic must motivate a great movement, convincing reluctant leaders and reluctant governments and reluctant peoples themselves to effect the needed changes."

Five areas must be addressed, the document's authors say, laying out an agenda that does not differ substantially from that advocated by many environmentalists. First, we must identify and limit environmentally damaging activities in order to "control and protect the integrity of the earth's systems," they say. Next, we must better manage resources like energy, water, and soil. Third, it is crucial that we stabilize population, and fourth, that we "reduce and eventually eliminate poverty." Finally, the authors call for sexual equality, in particular granting women the right to control their own reproductive decisions. "No more than one or a few decades remain before the chance to avert the threats we now confront will be lost and the prospects for humanity immeasurably diminished," the scientists warn.

No less an authority than Edward O. Wilson, the eminent Harvard biologist and author, and one of the document's signatories, has admitted, "The truth is that we never conquered the world, never understood it; we only think we have control. We do not even know why we respond a certain way to other organisms and need them in diverse ways, so deeply. The prevailing myths concerning our predatory actions toward each other and the environment are obsolete, unreliable and destructive." Ever the scientist, Wilson adds, "The more the mind is fathomed in its own right, as an organ of survival, the greater will be the reverence for life for purely rational reasons."

It is significant that so eminent a scientist as Wilson has identified a basic problem in our current approach and a compelling need for change. In that, he echoes the concerns of individuals from all walks of life who share this belief. Father Thomas Berry, a monk and "eco-theologian" who has eloquently described science as "the Yoga of the West," has written that "this endeavor over the past three centuries might be considered among the most sustained meditations on the universe carried out by any cultural tradition." He nonetheless states unequivocally that "the earth entire and the human community are bound in a single destiny, and that destiny just now has a disintegrating aspect."

Like many contemporary ecospiritual writers, Berry believes that science must turn its Gaze upon itself and assume a level of responsibility commensurate with its knowledge. "What we need, what we are ultimately groping toward, is the sensitivity required to understand and respond to the psychic energies deep in the very structure of reality itself. Our knowledge and control of the environment is not absolute knowledge or absolute control. It is a cooperative understanding and response to forces that will bring about a proper unfolding of the earth process if we do not ourselves obstruct or distort those forces that seek their proper expression. I suggest that this is the ultimate lesson in physics, biology and all the sciences, as it is the ultimate wisdom of tribal peoples and the fundamental teaching of all great civilizations. . . . If responded to properly with our new knowledge and our new competencies, these forces will find their integral expression in the spontaneities of the new ecological age. To assist in bringing this about is the present task of the human community."

Wilson and other scientists might dispute Berry's belief in "forces seeking their proper expression" (although if the force were defined as natural selection, perhaps not), but near the end of *Biophilia*, Wilson identifies the same paradox contemplated by both Bill Russell and Thomas Berry. "Natural philosophy has brought into clear relief the following paradox of human existence," he says. "The drive toward perpetual expansion—or personal freedom—is basic to the human spirit. But to sustain it we need the most delicate, knowing stewardship of the living world that can be devised. Expansion and stewardship may appear at first to be conflicting goals, but they are not. The depth of the conservation ethic will be measured by the extent to which each of the two approaches to nature is used to reshape and reinforce the other."

It may well be that the war between science and animal protection has been in some ways a microcosm of a larger conflict between the human desire to use nature and its resources to improve human life and the painful recognition that in doing so we run the risk of destroying the delicate symbiosis between ourselves and the natural world. That our anxiety-driven attempts to dominate and control not only increase an alienation we have not yet ceased to mourn but presently imperil our own survival seems self-evident. But as Thomas Berry points out "the industrial context in which we presently function cannot be changed significantly in the immediate future. Our immediate survival is bound up in this context, with its beneficial as well as its destructive aspects. What is needed, however, is a comprehensive change in the control and direction of the energies available to us."

Berry believes that science itself will play an integral role in resolving the current crisis: "We can do nothing adequate toward human survival or toward the healing of the planet without our technologies." Luc Ferry says the same thing in *The New Ecological Order*. Despite his denunciation of the totalitarianism implicit in certain forms of radical environmentalism, he recognizes the very real threat these movements have risen to combat and agrees that proceeding with a "laissez-faire liberal attitude is nonetheless insane. That it will ultimately be by means of advancements in science and technology that we manage one day to resolve the questions raised by environmental ethics is more than likely. Yet to imagine that these solutions will appear on their own, as if part of the natural evolution of things, is childish."

Then, too, as Rosemary Reuther points out in *Gaia and God*, "if dominating and destructive relations to the earth are interrelated with gender, class and racial domination, then a healed relation to the earth cannot come about simply through technological 'fixes.'" Reuther argues that instead we need "new visions of how life should be more just and caring." Like the ecofeminist authors quoted earlier, she views our essential problem as a crisis of conscience in which the human species is finally being forced to confront the consequences of our androcentric behavior. The "cure" for this fundamentally spiritual ailment cannot then be simply technological. Instead, it must combine the material with the spiritual, healing the schism that in itself has created the crisis. "The human capacity for ethical reason is not rootless in the universe," she writes, expressing a perspective that has long been absent from public discussions about the means and ends of science and technology. "Consciousness and altruistic care are qualities that have some reflection in other animals, and indeed are often poorly developed in our own species. To believe in the divine means to believe that those qualities in ourselves are rooted in and respond to the life power from which the universe itself arises."

The success of the more progressive elements of the research and animal protection communities in crafting a creative synthesis that has proven effective as a vehicle for change in the philosophy and practice of animal experimentation may offer a model useful in this larger debate. This process has taken different forms in the United States and various countries in Western Europe, as each nation has sought and implemented reform in a culturally distinct fashion. In the Netherlands, dialogue between government, industry, and three antivivisection societies composing the Alternatives to Animal Experiments Platform, together with the mandatory education of scientists in the ethics of animal experimentation, has fostered a system that has been called the most progressive in the world. Great Britain, building on the historic 1876 Cruelty to Animals Act system of licensing and regulation, adopted an even more stringent system of governmental oversight in the Animals (Scientific Procedures) Act of 1986, which mandates a cost-benefit analysis for each experiment as a prerequisite to approval. In the United States, self-regulation by the research community remains the favored model, with institutional animal care and use committees and accreditation by independent organizations such as the American Association for the Accreditation of Laboratory Animal Care, together with regulatory oversight exercised by the USDA, maintaining maximum independence for the individual researcher. Each of these strategies is an attempt to come to terms with the ethical complexity of animal experimentation and to craft political compromises using the democratic process and institutions of the individual nation.

The results may fall short of the expectations of those who wish for a revolution in human-animal relations. However, as Michel Foucault once pointed out in an interview, "We know from experience that the claim to escape from the system of contemporary reality so as to produce the overall programs of another society, of another way of thinking, another culture, another vision of the world, has led only to the return of the most dangerous traditions. I prefer the very specific transformations that have proved to be possible in the last twenty years in a certain number of areas that concern our ways of being and thinking. . . . I prefer even these partial transformations that have been made in the correlation of historical analysis and practical attitude, to the programs for a new man that the worst political systems have repeated throughout the twentieth century."

—ELEVEN—

PARTIAL
TRANSFORMATIONS

*Public policy, of course, is not made in a vacuum and is never
simply a product of moral principles, however valid one may
think them to be.*

—Robert Garner,
Animals, Politics and Morality

THE NETHERLANDS

WITH 15.8 MILLION PEOPLE living on only about sixteen
thousand square miles of land, the Netherlands is home
to a people skilled at accommodation, tact, and compromise.
"The Germans have a word to describe the Dutch temperament,"
said Otto Postma, CEO of Pharming Healthcare Products, a
Rockville, Maryland, biotech firm whose parent company,
Pharming Healthcare Inc., is based in the Netherlands. "The
word means 'two souls in one bosom.'"

Postma, a Dutchman, tells the story to explain to his audience
at the Second Annual Congress on Mammalian Cloning why the
Netherlands biotechnology legislation is so paradoxical—on the
surface quite stringent, reflecting Dutch concerns about animal
protection, but also extremely flexible, with saving loopholes for
research. In effect, scientists may breed animals that might be
potential donor animals in the Netherlands and conduct experiments on animals that have actually been produced via cloning

techniques there, but scientists may not carry out the actual transfer of genetic material within the country's borders. Some scientists in other countries have dubbed this policy hypocritical, and Gerald van Beynum, vice president for corporate strategy and communication and former chief scientific officer at Pharming Healthcare in the Netherlands, admits that "it is indeed a kind of crazy situation."

In his presentation at the cloning congress, Otto Postma sought to explain the Dutch tendency to combine idealism with a kind of brutal realism as a natural outcome of the country's history. Calvinism, particularly its belief in predestination and encouragement of "an acute sense of humility in its followers," combined with a love of commerce, led to the development of a split consciousness in the Dutch, Postma said, "a successful marriage of opposing viewpoints." Postma pointed out that in the sixteenth century, the Dutch East India Company, "the world's first multinational corporation," he joked, began slave trading—an ethically problematic enterprise it carried out by seeking biblical justification for the practice. Today the Dutch ability to tolerate high levels of ethical dissonance is legendary. Euthanasia, prostitution, and marijuana use are just three of the difficult social issues the Dutch have addressed in a characteristically double-handed manner, permitting these practices without explicitly condoning them. This approach is called by the Dutch *gedogen,* meaning "by law not really allowed, but for the time being you can do it under certain circumstances," said van Beynum.

This ability to moderate extreme viewpoints and to create a kind of morally ambiguous compromise has given the Dutch a reputation for being one of the most progressive countries in the world when it comes to resolving the cultural conflict surrounding animal experimentation. "The essence of any effective system is dialogue between all concerned parties. This is amply illustrated in the Netherlands where a new system of control based on co-operation between veterinarians in laboratories, institutional committees and a small central inspectorate seems to be working well," commented alternatives advocate Judith Hampson. "The Netherlands is perhaps the most advanced country in terms of addressing the moral issue of animal experimentation."

Indeed, when Wade Roush, a reporter for *Science* who was researching a story on the development of alternatives in 1996, attempted to locate a Dutch scientist critical of the adoption of the three R's as national policy, he couldn't find one. Although the Dutch passed a law in 1977 requiring scientists to use nonanimal methods whenever possible to replace animal experiments, which was followed by a dramatic drop in the number of animal experiments, from 1.6 million in 1978 to 673,000 in 1994, scientists in the Netherlands interviewed by Roush did not appear to feel that academic freedom or scientific integrity had been compromised. "People are much more strict in planning their experiments, but I don't think that animal work suffers from that," geneticist Richard Bootsma—identified as a "critic" by other scientists—told Roush. Roush found this relative lack of hostility from animal researchers matched by the civility of their adversaries. Geoffrey Decker, the head of PETA in the Netherlands, explained the organization's cordial relations with its adversaries succinctly: "In Holland we have a tradition of compromising."

Paul de Greeve, senior public health officer of the Ministry of Health, Welfare, and Sport in the Netherlands said much the same thing during the Second World Congress on Animal Use and Alternatives in the Life Sciences, held in Utrecht in 1996. "We don't always agree, but we don't fight anymore," said de Greeve, describing the relationship between the country's scientists and animal protectionists. De Greeve is a member of the Netherlands Alternatives to Animal Experiments Platform, funded by the Dutch Parliament in 1988 to provide a forum for the discussion of diverse viewpoints and to dispense grants for research on alternatives. The platform is composed of representatives from industry, government, and antivivisection and animal welfare groups. There is no equivalent government group, representing such a broad range of perspectives, in either the United States or the United Kingdom.

However, outside observers seem less willing than the Dutch to believe that this strategy of dialogue and compromise is specifically Dutch—or that it is as effective as advertised. "The consensus-building approach is typically Danish and has influenced many of the Scandinavian and Low countries," said Mark Matfield, director of the European Biomedical Research Association. "But then the Netherlands has also been influenced by the Flemish dictatorial style. The Netherlands also has one of the most dogmatic and effective animal rights movements in Europe— much closer to the U.K. and German movements than many other EU countries."

When it comes to biotechnology, the law passed on April 1, 1997, has not shut down animal experiments but has made the process far more frustrating and difficult. Each project, whether a single experiment or a group of experiments, must be individually approved by the Ministry of Agriculture—a process that can take from six months to more than a year. When the law went into effect, Pharming Healthcare already had a number of projects in the pipeline. The minister granted retroactive permits for these projects in January 1998. The company's scientists, however, were not granted permits to carry out cloning experiments in the Netherlands. Consequently, that work is now being carried out in the United States by the company's corporate partner Infigen Inc., based in Madison, Wisconsin. "The pregnancies are in Wisconsin, and the animals will grow up in the U.S.," van Beynum said.

The decision to deny Pharming Healthcare a permit to carry out nuclear transfer experiments within the borders of the Netherlands was "a political decision," van Beynum said, and though it will not affect the company's overall operations, since the work can be done elsewhere, it remains frustrating. Nonetheless, van Beynum himself seems remarkably free of the kind of bitterness or resentment that has often characterized U.S. researchers' response to increased government oversight and regulation. While admitting that university researchers in the Netherlands were angry about the restrictive nature of the new biotechnology law, he appreciates the constraints that temper the decisions of the Minister of Agriculture. "Through this very bureaucratic and lengthy process, the government is trying to avoid any risk," he said. "The minister is under pressure from Parliament. Because of that, he will always make the safe decision," even, said van Beynum, when that decision is not in the best interest of science. "Under the circumstances, he will always take the safe side."

As the new biotechnology legislation indicates, support for animal protection and environmental issues comes from the highest levels of government in the Netherlands. The Second World Congress on Animal Use and Alternatives in the Life Sciences, held in October 1996, was opened by Erica Terpstra, the state secretary for Health, Welfare, and Sport in the Netherlands, a position equivalent to the U.S. secretary of Health and Human Services. In her introductory remarks at the congress, attended by more than a thousand scientists from over forty countries, Terpstra said, "The role of governments with regard to animal experimentation is a very delicate and complicated one. On the one hand, the government's role is that of guardian of public health, and to protect this most fundamental of interests, it is necessary that animal experiments be carried out on the government's behalf. On the other hand, governments are bound by the increasingly apparent moral obligation to protect all living creatures. In other words, they must practice a responsible stewardship."

On February 5, 1997, the Dutch Parliament exercised what it believed to be responsible stewardship by inserting the concept of "intrinsic value" into its revision of the 1977 Experiments on Animals Act. Although this concept lends itself to many interpretations, it inarguably constitutes a fundamental acknowledgement of the moral status of animals. Commenting on the new Act, the Netherlands Center for Alternatives said: "With regard to laboratory animals, this [this concept of intrinsic value] implies that they are not to be reduced to mere instruments, but must be considered as sentient beings. They have a value of their own, independent of their utility value for humans."

The 1997 additions to the Act also banned animal experiments for cosmetics testing and provided for the establishment of ethical review committees to supplement the system of licensing already in effect. In the Netherlands, heads of research institutions are licensed by the government to perform animal experiments and are then responsible for ensuring that all animal technicians and researchers in the institution are adequately qualified and trained. The new ethical review committees, which must include at least seven members "in equal numbers experts in the field of animal experiments, experts in the field of alternative methods, experts in the field of animal welfare and protection, and experts in the field of ethical assessment," review individual experiments with the power to ban or approve the procedures. A negative recommendation can only be overridden by an appeal to the Central Animal Experiments Committee at the Veterinary Health Inspectorate.

One aspect of the Netherlands system that nearly everyone agrees has made a tremendous difference in the approach of scientists is education. All graduate students planning to perform animal experiments are required to attend a three-week course developed by scientists at the University of Utrecht on the ethical issues surrounding animal experimentation and the three R's—replacement, reduction, and refinement. More than three thousand Dutch graduate students and Ph.D. scientists have completed the course since 1986, according to University of Utrecht professor L.F.M. van Zutphen, who helped create the curriculum.

"The three-week course was developed in 1985 and organized for the first time in September 1986. In 1983, I was appointed as a professor in laboratory animal science at the Veterinary Faculty of Utrecht, and I was allowed to establish and lead a department which at that time had a staff of approximately ten persons," said van Zutphen. "This group decided to adopt the three R's as principles guiding our activities in both research and education (which was, at that time, not a matter of course). The course, which I developed in close collaboration with the members of our department and with feedback from representatives of the government, the scientific community, and the animal welfare organizations, was made compulsory by law in 1986 for 'new' scientists who are involved in the design or performance of animal experiments.

"In the beginning, the course was organized in Utrecht only, but now there are seven universities offering this course," he added. "All courses have the same 'format' and are centrally coordinated from the chair of laboratory animal science in Utrecht. The objective of the course is to contribute both to the welfare of animals and to the quality of research. Much emphasis is put on the student's attitude toward animals. Ethical aspects, alternatives, discussions with representatives from animal welfare organizations are as important as the 'technical' aspects of designing and performing animal experiments."

A small number of foreign students have also been trained by van Zutphen and his colleagues. "Once a year, a course is organized for foreign students," he said. "This 'international' course started in 1993. Since then, more than one hundred students from forty countries have participated. My personal feeling is that the course has a great impact on the scientists' attitude toward animals. They are more sensitive to the issue of animal welfare and are more willing to see possibilities to replace or reduce animal use, to seek methods and procedures that reduce suffering, or to contribute to the welfare of animals."

UNITED KINGDOM

DESPITE A LONG legislative history, animal protection remains a politically risky venture in Britain, particularly when it relates to animal experimentation. Although both Conservative and Labor governments officially support animal protection goals, when it comes time to stop talking and draft legislation, or even set up committees to investigate the possibility of crafting new laws and regulations, there are few volunteers for the job. "Animal experimentation is considered something of a poisoned chalice, politically speaking," said Mark Matfield, executive director of the Research Defence Society (RDS) in Britain. "When a new government allocates responsibilities to ministers, it always gets passed down to the most junior minister, the one who can't say no. Many politicians see it as a big headache, a no-win situation, in which you can never please everyone."

In addition to his RDS post, Matfield serves as director of the European Biomedical Research Association, giving him a unique perspective on the cultural aspects of the issue. The Research Defence Society is a venerable institution, founded in 1908 to counter the growing influence of the antivivisectionists and to disseminate information about the value of animal research to a suspicious public. EBRA, by contrast, was founded in 1994, reflecting the much later realization by scientific communities in other European nations that legislation in this area was important and that they needed to take an active role in the debate. "There are enormous differences across Europe in national attitudes, not just regarding animal welfare, but in attitudes toward regulation per se in the broadest social contract sense," said Matfield.

As director of EBRA, Matfield sometimes finds himself representing scientists in one country who are fighting legislation that scientists in another country in his constituency accept without complaint. This situation can be partially explained by geography, Matfield said. "There is a north-south divide in Europe, with attitudes toward animal experimentation more relaxed in the south," he commented, adding that in representing such a broad constituency, it is important not to view such diversity as a problem, but as a strength. "We must make a virtue [of the differing attitudes], rather than fight them, and try to benefit from the diversity, open to the concerns, approaches, and attitudes of individuals from different national backgrounds."

Matfield finds a wide gulf, however, between attitudes toward regulation in some parts of Europe and those in the United States, where "the legacy of a frontier mentality has resulted in an emphasis on individual freedoms," he said. There is a particularly acute disconnection between attitudes among scientists in the United States and those in the United Kingdom, he noted. "Our approach in Britain is that regulation is right and proper. We feel safe and happy when regulated. Yes, scientists complain, but basically they feel much more comfortable because [animal experimentation] is regulated." This may be partially because of a belief among scientists that "I can't do wrong because the regulations won't let me do wrong, as long as I'm playing by the rules," he said.

The rules in Britain were tightened in 1986 via the Animals (Scientific Procedures) Act, successor to the 1876 Cruelty to Animals Act. Under the 1876 Act, researchers were licensed to perform animal experiments, while "the mainstay of the 1986 system is the project license," Matfield said, which mandates that individual animal protocols also be approved by Home Office inspectors, who must be either physicians or veterinarians and are often also Ph.D.-level scientists. "They are usually at the level of a full professor when hired by the Home Office," Matfield explained. The 1986 Act also sets standards for animal care and housing and mandates a cost-benefit analysis for projects, requiring the investigator to consider whether the information to be gained from the experiment exceeds its cost in animal life and suffering. In addition, investigators are required to determine whether or not alternatives to the whole animal model exist for the particular question under study. Licenses can be revoked or fines levied if an individual or institution fails to comply with the Act.

"The Animals (Scientific Procedures) Act 1986 provides specific authority for specific work, rather than simply a list of what may be done irrespective of why and in what context," notes a chapter on the legislation in a text written for graduate students, *Laboratory Animals: An Introduction for Experimenters*. This goal is achieved by a dual system of licensing. Personal licenses, which permit qualified individuals to engage in experimentation, list acceptable actions or techniques, while project licenses, which must be secured for each and every protocol, "describe a specified work programme in sufficient detail for its implications to be addressed," according to C. B. Hart, author of the chapter.

A key aspect of this assessment is the cost-benefit analysis, which computes the potential benefit of the work versus its adverse effects on the experimental animal or animals. "Potential researchers must consider in depth the need for the work they propose and think out the details of its methodology as far as possible," says Hart. "They must also consider carefully the need to use protected animals or whether 'alternative' methods can be used instead." Although some animal protectionists have leveled the same criticisms at this cost-benefit analysis as they have at the system of animal care and use committees in the United States, notably that it is a pro forma exercise that does not seriously affect the 'business as usual' climate in laboratories, the analysis does provide a vehicle for review. "The structure is of benefit to the inspectors and, it should be added, to the animals, because there is rational input and discussion before the work is authorised, giving the opportunity for the concept and the details to be modified in the animals' favour," says Hart. "There is a framework to the programme, which enables there to be proper monitoring and structured reviews of progress."

The rules on animal experimentation in the United Kingdom are very clear, as Mark Matfield pointed out, and the resulting lack of ambiguity produces a somewhat unforgiving attitude toward those who violate them. The case of Huntingdon Life Sciences, a contract testing laboratory infiltrated by a journalist sympathetic to animal rights in 1997, is instructive, Matfield said. The journalist videotaped an incident in which laboratory employees clearly mistreated a beagle, and when the tapes were made public, the government took swift action, removing the personal licenses of the individuals concerned and passing the individuals to the police for prosecution. After investigating the company, the Home Office set out a series of sixteen conditions it had to meet within six weeks to rectify the animal welfare and management deficiencies. Failure to meet the conditions would result in the removal of the institutional certificate, which would force the company to close.

"Despite meeting these conditions and keeping their certificate, over the next eighteen months, the company lost large amounts of business, and many were concerned that it would fold," Matfield noted, pointing out that it was mostly pharmaceutical industry customers who abandoned the company. "There was an ingrained feeling among scientists and business people that this company had transgressed in a very serious way." By contrast, Huntingdon's U.S. affiliate, target of a PETA videotape sting revealing mistreatment of primates in the company's Virginia testing facilities a few months after the U.K. incident, has largely recovered from the scandal, although Procter & Gamble, Huntingdon's largest client and co-target of the PETA sting, has dissolved its contracts with the company. "In the U.S., you feel that you are allowed to make a mistake and then take your punishment. Here in the U.K., people don't forget. You are perceived as tainted because you broke the rules."

"There is certainly an increased sympathy and awareness of animal rights issues" among both students and faculty in British universities, according to David Dewhurst, former professor of health science at Leeds Metropolitan University, now at the University of Edinburgh. Dewhurst, who has created more than forty computer-based learning programs for use in university science classes, states that, "the number of animal labs has been drastically reduced." Matfield of the RDS points out that although the use of animals for educational purposes is written into the 1986 Animals (Scientific Procedures) Act, the use of animals to gain manual skills—for example, to learn surgical techniques—is banned, though a few exceptions have been permitted. Consequently, the use of animal experiments in teaching settings has been eliminated or greatly curtailed following implementation of the Act.

David Dewhurst said that his own university discontinued live-animal teaching labs after the passage of the 1986 law. "It was a purely economic decision," he admitted, noting that large sums of money would have been necessary to bring the animal housing facilities up to the standards required by the Act and that the university simply couldn't afford to comply. "That was one of the most tangible impacts of the Act," Dewhurst said, adding that many universities closed their animal facilities at that time for the same reason. Even in universities that have retained animal labs, medical science courses offer both 'wet' and 'dry' practical groups, with the former using animals and the latter not, according to Mark Matfield. "Some places even have 'dry' groups who do not even use animal blood, let alone tissue or live animals," he remarked.

Even so, David Dewhurst feels that there is some resistance to the three R's explicitly articulated as such among British scientists. Talking about the computer simulations he has developed in the context of replacing animal experiments can be counterproductive, he has found, and he now avoids talking about replacement. "Instead, I talk about the educational benefits of the programs—that students get a better quality experience using the computer programs." Noting that in Britain today, animal experiments for educational purposes make up less than 1 percent of the total number of experiments carried out in the country, a statistic that is consistent with those of other European nations, Dewhurst said that this downward trend continues, despite the fact that enrollment at British universities has doubled over the past decade. Although he believes that the driving force is economic, as increased regulation has driven up the cost of animal experimentation, he said that faculty response to student concern about experimentation has also played a part. "Twenty years ago, a student would be pitched out if he didn't want to experiment. Today's faculty are much more sympathetic," Dewhurst observed, and will usually work with a student to find alternatives.

At the Second World Congress on Animal Use and Alternatives in the Life Sciences, held in the Netherlands in 1996, political scientist Robert Garner of the University of Leicester delivered a lecture titled "Animal Experimentation and Pluralist Politics." Discussing some of the ideas presented in his 1993 book, *Animals, Politics and Morality*, Garner noted that a political consensus on animal protection similar to that observed in the Netherlands is developing in the United Kingdom and the United States as well, albeit on a more limited scale. "The state's role, largely confirmed by the evidence from Britain and the United States, is to balance the demands of competing legitimate interests and to try and promote consensus. To expect anything else, as advocates on both sides of the debate over animal experimentation frequently do, is unrealistic."

Garner pointed to evidence that public concern about laboratory animals has been met by sustained legislative action in both Britain and the United States. However, "legislative controls on the use of animals are not as stringent in the U.S. as they are in Britain," he said at the world congress in Utrecht. "This can partly be explained by the climate against government regulation, the influence of the research community and its allies within the National Institutes of Health." The resistance of U.S. researchers to government control is legendary among European scientists, particularly British scientists accustomed to more than a century of oversight. Despite the high level of regulation in Britain and the relative independence of American researchers, both countries have been confronted by an aggressive and at times violent animal protection movement over the past two decades.

In Britain, with its long legislative history of animal protection, the opposition has been particularly severe. This paradox is partially explained by class divisions, according to Mark Matfield of the RDS, who noted that the Animal Liberation Front in Britain grew out of the Hunt Saboteurs Association (HSA), a group founded in 1964, to disrupt the activities of largely upperclass hunters. Robert Garner traced the group's history in his book, noting that the British ALF was formed in 1976, "although its origins lie in a breakaway faction of the HSA dating back to the 1960's. Led by Ronnie Lee and Cliff Goodman, a group of activists became disillusioned with the law-abiding activities of the HSA and, in 1972, formed the Band of Mercy (a name ironically borrowed from the RSPCA's law-abiding youth organization in the nineteenth century). This group began by damaging the property of hunt participants and supporters but quickly moved on to other targets. Following his release in 1976 from a prison sentence imposed after he was caught breaking into an animal breeding center at Bicester, Lee formed the ALF, initially a group of about thirty."

Unlike the U.S. ALF, which has avoided acts of personal physical violence despite its threats, the British ALF and its splinter groups have copied some of the tactics of the Irish Republican Army and other terrorist groups, sending letter bombs to government ministers in 1982, claiming to have poisoned candy, cosmetics products, and soft drinks, and setting fires in department store fur departments in the 1980s. "The use of high explosive which severely damaged the Senate House at Bristol University in February 1989 represented a further disturbing escalation of the campaign and the final step towards fully-fledged terrorism occurred in 1990 when, in the space of two weeks, cars belonging to Margaret Baskerville (a veterinary surgeon working at the Porton Down chemical research establishment) and Max Headley (a Bristol University psychologist) were blown up by animal rights activists. Although neither was hurt, this was more by chance than design and in the latter case a thirteen-month-old baby was injured by the blast," Garner wrote.

In the United States, individual researchers have received death threats by mail or telephone and have endured extensive harassment with psychological costs, but as of January 2000, none has suffered physical harm. On the other hand, in Britain, activists themselves have been killed during direct action activities: in 1991, an eighteen-year-old man died during a hunt-saboteur protest, when he was dragged beneath a pickup truck he was attempting to jump from, and in 1995, a thirty-one-year-old woman was hit by a car during a veal crate protest in Coventry. The economic damage inflicted on laboratories by animal activists in Britain also exceeds the amount of damage in the United States, according to Garner. "It has been claimed, for instance, that between November 1982 and 1986, 2,000 separate actions caused £6 million worth of damage" (about $12 million in 1998 U.S. dollars), while in the U.S. between 1979 and 1990, the ALF was responsible for "seventy-five major actions, freeing over 3,000 animals and causing more than $4 million worth of damage," Garner said.

The U.S. research community took aggressive action against animal rights break-ins and vandalism of laboratories and other animal-related enterprises by pushing for the passage of the Farm Animal and Research Facilities Protection Act, passed by Congress in August 1992, which makes such acts federal crimes. No similar legislation has been passed in Britain, although Mark Matfield of the RDS (the intended recipient of a letter bomb that was intercepted by police before its arrival) said in 1998 that since the British intelligence community set up an animal rights task force, the ALF in the United Kingdom has been successfully infiltrated and effectively defanged.

That assumption was called into question, however, by events in 1999, including the kidnapping and branding of an investigative journalist who had produced a documentary titled "Inside the ALF." The reporter, Graham Hall, who had secretly taped activists discussing bomb production, arson, and harassment of enemies, was abducted on October 26, 1999. According to an article published in the London *Independent*, "The kidnappers put a hood over his head and pressed what appeared to be a gun against him. 'I thought that I was being taken to my execution, and I'm not ashamed to admit I was so frightened that I wet myself,' Mr. Hall said."

Instead, after several hours during which they kept him tied to a chair, Hall's captors pushed his head between his legs, lifted his shirt, and branded the letters ALF into his back with hot iron. After releasing him hours later, Hall's captors threatened him with death if he went to the police—a warning he disregarded. Hall, known to be an advocate for animals, was left with a four-inch by nine-inch scar on his back requiring plastic surgery. The story of Hall's abduction and branding was covered in both British and U.S. newspapers and was prominently featured on the Americans for Medical Progress Web site, together with an account of the mailing of threats and concealed razor blades to more than eighty scientists in the United States in November 1999.

In the aftermath of the razor blade incident, which is under investigation by the FBI, a number of animal protection groups attempted to distance themselves from such behavior. Ingrid Newkirk of PETA, however, upheld that organization's historic unwillingness to condemn the activities of the ALF and other groups willing to use violence and threats in support of the animal rights cause. Instead, she used the razor blade campaign as an opportunity to attack researchers. "This is a fight between money and decency and it's turned nasty," she told a reporter from the *Chronicle of Higher Education.* "When you see the resistance to basic humane treatment and to the acknowledgment of animals' social needs, I find it small wonder that the laboratories aren't all burning to the ground. If I had more guts, I'd light a match."

Incidents like the ALF branding and the mailing of razor blades hidden in letters are used as flags to rally the troops, to clearly demarcate "our team" from "theirs." This either-or approach to an enormously complex issue diverts attention from more serious efforts to reconcile the competing demands of science and animal protection and to build a platform for reform. Those efforts are clearly much further advanced in Europe than in the United States at present. In the United Kingdom, and in the Netherlands in particular, the desire to ban cosmetics testing on animals and to reduce the number of experiments performed for other purposes cuts across class lines and party affiliations. Such broad-based support gives the antivivisection movement the kind of political clout that it largely lacks in America, where the antics of PETA and its imitators seem to be eroding the once formidable support of Americans for animal advocacy.

"There is no doubt that Europeans have made remarkable strides in the development and acceptance of alternative methods, and have shown a willingness to incorporate the ideology into legislation, both at a national and at the community level," Gillian Griffin, a British scientist and alternatives researcher currently working in Canada, wrote in 1995. In an article for the newsletter of the Johns Hopkins Center for Alternatives to Animal Testing in which she explored differences in animal experimentation policy and practice in Europe and North America, Griffin concluded that the passage of Directive 86/609/EEC, ratified in 1986, fundamentally established the commitment of the European Union to the replacement, reduction, and refinement of animal use in science.

This law, the Council Directive on the Approximation of Laws, Regulations and Administrative Provisions of the Member States Regarding the Protection of Animals Used for Experimental and Other Scientific Purposes, has driven much of the reform in the European Union over the past fifteen years, with member states adopting similar legislation at the national level. Article 7(2) of the directive explicitly states that "an experiment should not be performed (on an animal) if another scientifically satisfactory method of obtaining the result sought, not entailing the use of an animal, is reasonably and practically available." EU directives like this one are drafted in Brussels and then implemented by member states, with each nation committed to bringing national legislation in line with community objectives. By the late nineties, all EU member states had some form of national legislation complying with the directive. However, as C. B. Hart pointed out in *Laboratory Animals*, "because of variations in detail and in the manner in which [the directive] is enforced between one country and another, it is difficult to be sure that the intention of harmonisation is matched by the reality."

Then, too, as Mark Matfield pointed out, animal use in science is driven by a number of factors that can either increase or decrease the need for experimentation. In the United Kingdom, where the government has compiled detailed statistics on the number of animal experiments by category since the passage of the 1876 Cruelty to Animals Act, the data show reductions of approximately 50 percent since 1974, Matfield said, with similar reductions in Germany, Switzerland, Sweden, and the Netherlands. The numbers are holding fairly steady now, he observed, although the use of transgenic animals is rising 20 to 30 percent every year in the United Kingdom. "I'm sure that's reflected in the statistics of every major country conducting research," he said. "We're looking at a shift in science, due to the available techniques. The techniques determine what questions can be asked."

Biotechnology is booming in Britain, and Matfield said that because of the high quality of the science, companies are willing to tolerate the "high hurdles of regulatory burden," which exceed those encountered in most other nations. However, he acknowledged, in international collaborative projects, the animal work is seldom done in Britain. The key question remains, What level of animal experimentation are people willing to accept to achieve socially sanctioned goals? "If society accepts the need to do transgenic work to discover the mechanisms of gene action in disease, that will obviously increase animal use," he said. The demand of customers for environmentally friendly products provides another example of a social good that will nonetheless increase animal use, even if only temporarily, as the development of new products leads to another round of ingredient testing to ensure that they are safe. "If the experiments are well designed, lead to useful outcomes, and minimize animal suffering," then society will have to accept some level of continued animal experimentation, Matfield said, if people want to realize the benefits of the research.

UNITED STATES

THE HUMANE Society of the United States—with more than 3.5 million members, the largest animal protection organization in the world—awards certificates each year to scientists who have made outstanding contributions toward the advancement of alternative methods in research, education, and testing. In 1998, the Russell and Burch Certificate of Recognition was given to two U.S. government employees who spend far more time in meetings than in laboratories: William Stokes and Neil L. Wilcox, both of whom are veterinarians, a group that has played an important role in improving standards of care for research animals in the United States over the past two decades.

William Stokes is director of the National Toxicology Program's Interagency Center for the Evaluation of Alternative Toxicological Methods, based at the National Institute of Environmental Health Sciences in Research Triangle Park, North Carolina. He also cochairs the Interagency Coordinating Committee on the Validation of Alternative Methods (ICCVAM). Established as an ad hoc committee in 1994 and as a standing committee in 1997, ICCVAM coordinates the federal government's involvement in the development, validation, and regulatory acceptance of alternative testing methods. Neil Wilcox received his award while serving as senior science policy officer with the Food and Drug Administration, a position he left in 1999 to become director of scientific and regulatory affairs for Gillette. Wilcox has a long history of involvement with alternatives, having chaired ICCVAM's predecessor, the Interagency Regulatory Alternatives Group, and helped form the FDA's Subcommittee on Toxicology, which reports to the agency's Scientific Advisory Board. The subcommittee was created to help the agency incorporate in vitro tests and other alternative approaches into regulatory protocols.

Stokes and Wilcox had long served as key figures in the federal government's efforts to grapple with the regulatory issues raised by the development of alternative testing methods. Wilcox, formerly a veterinarian in private practice, has been involved in the search for alternatives since 1990, when he was hired by Gerald Guest, former director of the Center for Veterinary Medicine at the FDA, to head the Office of Animal Care at the agency. Instructed by Guest to coordinate all animal-related matters within the agency, including the novel issue of alternatives, Wilcox was handed a tough assignment. In the early years, he says with characteristic understatement, "it was difficult to persuade the agency to engage the topic of alternatives." Like many U.S. researchers, regulatory scientists within the FDA saw the development of alternatives not as a pressing scientific issue but as a political problem.

Although by 1990 industry was in the midst of a substantial reevaluation of traditional testing practices and was funding research in the new disciplines of molecular toxicology and molecular epidemiology, regulatory agencies were for the most part wedded to the traditional whole-animal model. In the traditional toxicological model, batteries of tests measure both acute and chronic toxicity and localized effects in animal models. This type of toxicology relies on complex mathematical formulas to extrapolate toxic dosage levels in animals (often rodents) to human beings. Despite certain limitations, the whole-animal model had worked for more than forty years to ensure public health and safety—the FDA's primary concern—and few regulatory scientists were interested in meddling with success.

Nearly a decade later, the regulatory scene has changed dramatically, particularly at the FDA, which initiated an agency-wide review and revamping of testing practices in 1996. Wilcox stresses that the changes now in progress are science-based, that the agency's goal is "not to eliminate or even reduce the number of animals (although we endorse that goal) but to come up with tests that are more predictive, which will better determine the safety or hazard of a particular substance. Rather than coming up with a better animal model, we're interested in coming up with a better model, period." This effort will in the long run greatly reduce, if not eliminate, the use of animals, Wilcox believes.

"The technology that is emerging today looks at the cellular, sub-cellular, molecular, and genetic level, examining the effect of a particular chemical on a part of a molecular, a gene, or an enzyme. We're asking questions about the specific mechanism that causes the effect."

The changes in product testing practices that began in industry have thus spread to government. The gradual reshaping of toxicology and regulatory science now underway mirrors to a startling degree the process of paradigm shift described by historian Thomas Kuhn in his 1962 text, *The Structure of Scientific Revolutions*. According to Kuhn, "When the profession can no longer evade anomalies that subvert the existing tradition of scientific practice—then begin the extraordinary series of investigations that lead the profession at last to a new set of commitments, a new basis for the practice of science. The extraordinary episodes in which that shift of professional commitments occurs are the ones known as scientific revolutions."

Such radical retooling of a scientific discipline is not a painless process. Those who argue that in vitro tests can never truly replace whole-animal methods illustrate a concept that Thomas Kuhn termed incommensurability. "The proponents of competing paradigms tend to talk at cross-purposes—each paradigm is shown to satisfy more or less the criteria that it dictates for itself and to fall short of a few of those dictated by its opponent," wrote Kuhn. In regard to toxicology, critics of whole-animal methods claim that species variability invalidates the animal model, and proponents of traditional in vivo models insist that in vitro tests cannot predict the systemic effects of a toxic substance in a whole animal or human subject. As Neil Wilcox notes, many old-school toxicologists, trained by using whole-animal models, are loath to abandon a field in which they are recognized experts for a discipline in which the questions themselves are unfamiliar. Although some might argue that the shifts now underway in toxicology have been driven by animal welfare, not scientific considerations, the fact remains that the limitations of the whole-animal model have become increasingly apparent over the past fifteen years. By the 1980s, some scientists had begun to question whether certain kinds of rodent studies were really predictive of human response, particularly when compared with the powerful new tools being developed by molecular toxicologists. Although this new discipline is in its infancy and is not yet able to provide many answers, inarguably it has created an entirely new set of questions.

In *The Structure of Scientific Revolutions*, Kuhn pointed out that radical changes in a research program are rarely completed until the older generation of scientists dies off or retires and proponents of the new paradigm are installed in positions of power and influence. This seems a particularly apt description of current regulatory policy. One of the greatest problems in implementing the new approach is getting a broad range of "stakeholders" as they are known in policy parlance, to buy into the process of gradual change, Wilcox points out. "It's definitely a long-term process," he says, "and one of the problems is getting people to see the vision, the big picture. It's a tough road to go down, particularly now, when resources are so limited." Developing new tests is an expensive endeavor and will require the cooperation of government, industry, academia, and the public.

The role of the latter should not be underestimated, Wilcox said, pointing out that in Europe, tremendous public pressure has driven regulatory reform throughout the nineties. That pressure has been lacking in the United States in recent years, although the outcry over the Environmental Testing Agency's High Production Volume (HPV) testing program may serve to awaken public interest in the testing issue once again. On October 9, 1998, the Clinton administration, the Environmental Defense Fund, and the Chemical Manufacturers' Association jointly announced a six-year program to test 2,800 major industrial chemicals for their health and environmental effects. This announcement followed three studies that indicated that most industrial chemicals produced in high volume in the United States lack even the most rudimentary information on potential health effects.

In a 1997 report titled *Toxic Ignorance,* the Environmental Defense Fund concluded that for nearly 75 percent of the chemicals produced in high volume in the United States—chemicals ubiquitous in our home and work environments—basic toxicity tests either have not been conducted or the data from such tests are not publicly available. "In other words, the public cannot tell whether a large majority of the highest-use chemicals in the United States pose health hazards or not—much less how serious the risks might be, or whether those chemicals are actually under control. These include chemicals that we are likely to breathe or drink, that build up in our bodies, that are in consumer products, and that are being released from industrial facilities into our backyards and streets and forests and streams," the report stated.

Following the EDF's study, the U.S. Environmental Protection Agency and the Chemical Manufacturers' Association, an industry trade group, conducted their own literature reviews and obtained similar results. Each of these studies found that basic health effects information on most major industrial chemicals is not "publicly available." As a result, Vice President Al Gore, whose long-standing interest in environmental issues is well known, worked to create a cooperative agreement between EDF, EPA, and the CMA to accelerate testing of the nearly 2,800 industrial chemicals produced or used in the United States in volumes of more than one million pounds per year. Information on these chemicals will be posted on the EPA's Web site and in other media.

The benefits of the HPV program, one aspect of a larger project dubbed the Chemical Right-to-Know Initiative announced by Gore on Earth Day 1998, seem at first glance inarguable. Chemical manufacturers have been asked to voluntarily submit data on their HPV chemicals. Much of the necessary information, though not publicly available, is believed to exist in company files and industry databases. HPV chemicals not voluntarily reported on by industry will then be tested by the EPA, with the agency also pursuing testing of chemicals known to be toxic to some degree and also known to persist in the environment and "bioaccumulate" in the bodies of human beings and animals. Finally, and perhaps most significantly, HPV chemicals of particular concern to children's health will be the subject of more detailed and extensive testing in a separate Children's Health Test Rule. The entire process of data review and testing is scheduled to be completed by 2004 at a cost of approximately $500–$700 million, most of that cost borne by industry. The EPA will be responsible only for the cost of administering the program.

However, the HPV and Right-to-Know initiatives were immediately attacked by a number of animal protection organizations who denounced the methodologies to be used in the testing programs as both unnecessary and cruel. On February 24, 1999, PETA ran a full-page ad in *The New York Times* denouncing Gore and his support of the HPV and Right-to-Know testing programs. "If you think Al Gore is an environmentalist, think again. First he was caught in a secret White House memo supporting commercial whaling. Now he's pushing a government program that will kill millions of birds, fish, rabbits and other animals in useless and painful experiments," the ad read. "It may sound like something to vote for but it's really just bad news for the environment, the animals and public health. Animal tests are so unreliable that they can actually clear chemicals we already know to be harmful to humans. Or, when animals die, companies can claim the results don't apply. Either way, government action will be delayed for years. Modern, reliable, non-animal tests are available but are being ignored."

The criticism of the animal advocates produced results. On June 17, 1999, the House of Representatives' Subcommittee on Energy and the Environment held a first hearing on the HPV program, three months after Science Committee chairman F. James Sensenbrenner sent a letter to EPA director Carol M. Browner, requesting the agency's response to ten probing questions on the methodological and scientific assumptions underlying the program. Susan Wayland, acting assistant administrator, responded that the EPA would provide "a detailed response to all of your questions within the next two weeks." Two months later, the committee was still waiting for the EPA's reply, which was delivered less than twenty-four hours before the start of the hearing.

In his opening statement, Ken Calvert, chair of the House energy and environment subcommittee, expressed his dissatisfaction with the EPA and its planned strategy in the strongest terms, questioning both the scientific assumptions on which the program is based and the methodologies that will be used to assess toxicity. "Significant doubt exists about the validity of animal testing results when applied to humans," he said. "I would like to hear about some of the humane alternatives to animal testing." Calvert concluded his remarks by noting that "it might be better to go back to the drawing board on the HPV program; spend a little bit more time and apply some well-considered sound science to design a better chemical testing program."

Clearly influenced by the perspective expressed by the program's most vocal critics, the animal protection community, Calvert briefly reiterated some of the points that were later repeated in the testimony of two of the hearing's three witnesses. Of the three witnesses asked to testify before the committee, two represented animal protection groups—Neal Barnard, president of the Physicians' Committee for Responsible Medicine, an antivivisection organization, and Jessica Sandler, an industrial hygienist representing a consortium of nine animal protection organizations, including PETA, In Defense of Animals, Earth Island Institute, and the Animal Legal Defense Fund. The third witness, William Saunders, director of the Office of Pollution Prevention and Toxics of the EPA, was outnumbered and increasingly outgunned as the committee members joined the animal protection groups in criticizing almost every aspect of the HPV program. Saunders attempted to answer the objections but appeared increasingly besieged and not always particularly well prepared. By the end of the hearing, it seemed clear that the animal protection groups had won the first round and that the agency would be asked to undertake further review and modifications of its program.

It would be easy to file objections to the HPV program under "animal rights" and use that controversial label to dismiss concerns about the program—easy but wrong. There are legitimate scientific and policy issues at stake in the HPV debate, and these exist apart from the ethical concerns expressed by those who object to the human use of animals as test subjects. Nonetheless, it is true that many of these objections were first raised by the program's animal protection critics, and just as over twenty years ago, science, policy, and ethics combined to produce a revolution in cosmetics safety testing, so today animal rights pressure has worked as a catalyst to stimulate scientific review and consequent modification of the EPA's chemical testing program.

Although it is not well known, an EPA scientist named Richard Hill, whose involvement with the alternatives issue runs parallel to that of the FDA's Neil Wilcox, was also to be awarded HSUS's Russell and Burch Award in November 1998, along with Wilcox and Stokes. Like Wilcox, Hill was a founding member of the Interagency Regulatory Alternatives Group and an early advocate of alternatives within his agency. But unlike Wilcox, who was able to parley his expertise into a broad and deep review of testing protocols within the FDA, Hill did not receive much support or backing from top levels of management at the EPA, leaving the agency vulnerable to exactly the kind of trouble that erupted when it announced the HPV testing program. Hill declined the HSUS award, "for personal reasons," he said.

OVER THE PAST two decades, U.S. federal regulations on the care and use of laboratory animals have been gradually strengthened. The 1966 Laboratory Animal Welfare Act (LAWA), although perceived as a victory for animal protection by its supporters, did not really affect the care or use of animals in laboratories. Successive revisions and amendments to the Act have changed that, and though it is still perceived as a "paper tiger" by many critics, the legislation has nonetheless provided a mechanism for review of animal research practices. The Animal Welfare Act, as it is now called, is enforced by the United States Department of Agriculture's Animal and Plant Health Inspection Service through its Regulatory Enforcement and Animal Care (REAC) units. The REAC units work together to implement AWA through inspections of laboratories, compilation of statistics on animal use from institutional reports, and sanctioning of laboratories found in violation of the Act. However, much of the responsibility for actual in-house implementation of federal regulations falls on laboratory animal veterinarians.

"Historically, there's been a rough relationship between researchers and veterinarians," Neil Wilcox said. "It's been difficult for veterinarians to capture an adequate amount of control because researchers feel that they know what they need to do and do it." Noting that part of the job of a laboratory animal veterinarian is ensuring "that the IACUC is adequately reviewing protocols, and assisting the IACUC in implementing the three R's—particularly refinement and reduction," Wilcox said that the responsibility for the third R, replacement, falls to the investigator and that "the new databases are hopefully providing conduits for researchers to ensure that they have adequately researched alternatives."

Whether characterized as alternatives or "complements," the preferred term within the research community, the three R's are increasingly used in laboratories today, even by those scientists who do not consider themselves proponents of a three R's approach. The NIH Revitalization Act of 1993, which instructed the director of the agency to prepare a plan to conduct or support research that reduces, refines, or replaces the use of animals, establish the validity of these methods, encourage their acceptance, and train scientists in their use, has provided a legislative mandate for the three R's approach in the United States. Although Congress did not support this mandate with funding earmarked for a targeted research program, the Act led to the creation of the Interagency Coordinating Committee on the Validation of Alternative Methods (ICCVAM).

"ICCVAM has become exactly the focal point we needed," said Neil Wilcox of the FDA, "by providing a mechanism for a formal review of each new method." Although ICCVAM addresses testing and not basic research, Wilcox believes that the "understanding and sensitivity" implicit in the three R's approach has penetrated the entire research community. "*Alternatives* is no longer a dirty word," he says. "Some people may continue to deny the value of the three R's, but they are in the minority."

Verbal support for the principles of replacement, reduction, and refinement appears to be spreading throughout the biomedical research establishment. Louis Sibal, director of the Office of Laboratory Animal Research at the National Institutes of Health, said that "NIH does officially support the three R's, and is pretty much footing the bill for ICCVAM." Both law (the Animal Welfare Act) and policy (Public Health Service Guidelines) implicitly mandate a three R's approach, "so we do support them, but in a quiet way," he said. Sibal admitted that the U.S. approach can fairly be described as self-regulation. "But it works," he maintained. "There will always be groups who slack a bit, but with the oversight from OPRR [the NIH's Office of Protection from Research Risks] and the USDA, the system works. We make people justify what they're doing, through the process of peer review." Noting that in the United States, "We have 2,000 institutions doing animal research," Sibal pointed out that "you can't always be there" and therefore it is up to researchers themselves to comply with federal policy on animal experimentation.

The United States remains the largest user of laboratory animals in the world. According to a table published in *The Encyclopedia of Animal Rights and Animal Welfare* in 1998, the top six users of laboratory animals were the United States (13,955,000 a year), Japan (12,236,000), France (3,646,000), the United Kingdom (2,842,000), Germany (2,080,000), and Canada (2,042,000). The next eleven countries use less than 1,000,000 animals a year. The total number of animals used annually in biomedical research, testing, and education worldwide has been estimated to be over 41 million, though that figure is difficult to confirm because the kind of detailed statistics compiled in the United Kingdom and the Netherlands are not kept in most other nations.

The U.S. numbers are much higher than those in the European nations, according to Andrew Rowan, senior vice president for Research, Education, and International Issues at the Humane Society of the United States, because in the U.S. "there is a $40 billion research industry," six to eight times larger than the $5–$10 billion industry in the United Kingdom and other European nations. The higher number of animals used in experiments in the United States reflects this difference in proportion. However, Rowan also noted that greater resistance to reevaluating the place of animals in biomedical research in the United States, a resistance that he feels is reflected in the number, can be attributed to broad-based sociopolitical and cultural factors. "Europeans have grown up in societies more willing to accept social controls on behavior," he said, "while the U.S. developed as a country where individual rights are paramount. Laws and regulations constraining the rights of individuals are frowned upon here, while in Europe it is recognized that in order to live compatibly, it is sometimes necessary to sacrifice some of one's individual preferences."

Rowan and the HSUS recently initiated a four-part program to improve the alleviation of pain and distress in laboratory animals. An expert report on pain and distress will be used as a tool for outreach to animal care and use committees. Rowan himself is working with the USDA to revise its pain classification guidelines. Finally, HSUS is actively seeking public (NIH) and private (corporate) funding for a study of pain and distress in laboratory animals. These efforts have so far met with a disappointing response, according to Rowan. HSUS has sent out three mailings to more than 1,800 IACUCs, describing the program and inviting IACUCs to participate in the effort. The first mailing garnered only two responses, the second ten, and the third (which included a postage-paid return envelope) sixty-nine responses. Of that number, only fifteen IACUCs expressed an interest in participating in the program. The scientific community, according to Rowan, remains "suspicious of our intentions and our motives." Having worked in the field of scientific animal protection for nearly two decades, Rowan feels that "there is more anger and resistance to change now than there was three to four years ago. Many researchers feel that enough concessions have been made and that the whole thing has been going on for too long." Rather than viewing efforts to improve the well-being of laboratory animals as an ongoing process, many feel that at a certain point these issues should be considered resolved.

That is not likely to happen, however, as there is now a small but persistent group of reformers working within the scientific community for change. Even among those groups most committed to the status quo, there has been a somewhat significant change in outlook over the past few decades. At a series of three workshops sponsored by the Scientists Center for Animal Welfare (SCAW) from 1992 to 1994, the former chief of the Institute for Laboratory Animal Resources (ILAR) of the National Research Council, Thomas Wolfle, described the evolution of attitudes he has witnessed in U.S. laboratories over the course of his career. "I can remember giving a talk thirty years ago about the fact that laboratory animals are not pets," he said. "They have numbers, not names. We treat them as important parts of our environment, but we do not warm up to them. I have come to realize that that attitude is opposed to everything that I now believe about the well-being of animals and the quality of the research."

Wolfle, who stepped down as chief of ILAR in 1997, focused on important changes in the training and status of the individuals most directly responsible for animal well-being—the laboratory technicians who provide direct care to research animals. "In a paper back in the 1960s, I referred not to technicians, but to 'dieners' " [from the German *Diener*, "servant"], Wolfle said. "The label subsequently changed from dieners to caretakers, and from caretakers to caregivers. I think we are trying to tell people as well as ourselves that they are important." In many respects, the technicians have the hardest job in the lab, Wolfle admitted. "We hire technicians for their compassion as well as their ability. . . ." We require that technicians spend time listening to and getting to know individual animals, because there are great individual differences. We encourage, through that process, the development of what we call a bond. We require them to kill the animal at the end of all that."

In recent years, some laboratories have begun to address issues of grief and moral conflict among laboratory personnel and to provide counseling for employees. Joseph S. Spinelli, director of the animal care facility at the University of California at San Francisco, has been a leader in this area, appointing nurse-psychologist Betty Carmack to provide therapeutic services for laboratory staff, including principal investigators (the scientists in charge of experiments), researchers, veterinarians, and technicians. Spinelli and Carmack discovered that researchers, too, experienced many of the feelings described by the technicians. "Some principal investigators reported to Dr. Carmack that their feelings were getting in the way of being able to use animals in research," said Spinelli. "This was especially true when they had to kill the animals. They felt like they were struggling with their core values." Although provocative and compelling, it may not be possible to generalize these admissions by a small group of scientists in one institution to the population of scientists at large. As Spinelli noted, these individuals volunteered to participate, and their willingness to do so indicates that they are particularly aware of their conflicting feelings about their work.

The vast majority of researchers tend not to discuss their feelings about using laboratory animals or about the psychological costs of public disapproval or animal rights harassment. Spinelli's paper describes a variety of coping mechanisms that scientists use to come to terms with the realities of laboratory life, including denial and rationalization, blaming authority, objectifying and distancing, and in rare cases, animal abuse. It is important to note that Spinelli does not object to animal research but that he is committed to ensuring the best possible care for laboratory animals and to helping laboratory staff confront and address their feelings about their work. By doing so, he says, they not only learn how better to cope with the inherent stress of the work but also develop better relationships with the animals in their care: "All of the coping mechanisms I have described exert a severe burden on humans and the animals with whom they interact, leading to job burnout, high turnover rates, reduced production, pathological self-directed behavior, and even animal abuse. I believe there are better ways to deal with unpleasant situations that arise in using animals for research, things we can do that will extract less psychologically from the humans and improve the quality of life for the animals."

Spinelli's approach is rare. Many laboratory workers express the same fear described by Joy Becker, a technician in Spinelli's lab, when she spoke at a meeting of the American Association for Laboratory Animal Science (AALAS) about the university's program. "I felt there a tremendous fear from participants that discussion would open up a Pandora's box of feelings and that people would not know how to deal with those feelings. I thought that it was time for AALAS members to take the initiative and demonstrate that we had a great amount of care and love for animals. I spoke about my personal experience going down the path of demons, as I called it, and looked at my own sadness in working with animals," she commented in a question-and-answer session at one of the SCAW workshops.

The "path of demons" described by Becker—confronting one's own ambivalence and pain about the emotional demands of animal research—has been described by other laboratory workers. Women scientists, veterinarians, and technicians are often particularly open in describing their struggle to balance their belief that the work is important and necessary with their emotional response to the animals who are subjects of experiments. For example, neuroscientist Candace Pert, who discovered the opiate receptor while working as a graduate student in the laboratory of Solomon Snyder at Johns Hopkins University in 1972, described the process of coming to terms with animal experimentation in her scientific autobiography, *Molecules of Emotion*. "At Bryn

Mawr, my early science training had been in the classroom of a Miss Oppenheimer, a fine teacher who almost threw me out of the department because of my stubborn, albeit principled refusal to kill a frog for dissection. There was some emotion in me that would not allow me to kill an animal. The thought of pulling apart a creature that I myself had just killed, no matter how marvelous its structure or incredible its fluids, made me sick to my stomach."

As a graduate student, Pert realized, however, that to do the work she loved she would have to overcome this obstacle. "I knew that I was going to have to desensitize myself if I was to succeed, and so I began the gradual process of rewiring my nervous system a good week in advance of my first day on the opiate receptor project. Each day I forced myself to stand a little closer to the door of the room where they did the killing. After a few days, I was able to stand in the doorway and watch as the animals were decapitated, a procedure done with a slick little guillotine, allowing the brain to be quickly scooped out and immersed in a cold liquid buffer that kept the neurons alive and nourished while freezing the internal chemistry. Then I killed one myself. My hands trembled and my heart pounded, but I forced myself to do it. It was so traumatic, I had to sit down after that first time to regain my composure."

As time went on, Pert was able to carry out this task when necessary, she says, but she was never able to do it automatically, without feeling. "There was always a ritual to it, an awareness that this was a sacrifice of life for life," she says. Like most scientists, Pert views the taking of animal lives as a necessary trade-off for potential cures to save human lives, and she was able to accept the exchange if it was carried out with respect for the animal and in a way that did not inflict suffering. "I made the choice I believed was the right one," she wrote—though obviously that choice was not made without a certain degree of discomfort and a willingness to suffer the agony of moral conflict.

"Regardless of how ethical you believe the use of laboratory animals is, you will encounter unpleasant situations and have negative feelings about them," Joseph Spinelli wrote. "Those feelings become more intense when one has grown fond of the animals. In fact, in such cases there will be feelings not only of guilt but of sadness as well." Blunt and brutal admissions like these are not popular in the research community, where words like *guilt* and *sadness* tend to be rejected as evidence of animal rights sympathies rather than as normal human responses to the demands of research. Arnold Arluke, professor of sociology and anthropology at Northeastern University, who has done field work in more than thirty-five laboratories, tells a funny story about the research community's discomfort with words like these, and with the feelings they describe.

In 1987, Arluke was invited for the first time to speak at a meeting of researchers about the results of his own investigations into the culture of science. He chose a provocative title for his talk, "The Experimenter's Guilt." "I was told that my choice of title was too controversial, even though the content of the talk was acceptable," Arluke says, "and that I should change *guilt* to something more palatable, such as *stress*." Soon after, the journal *Lab Animal*, a magazine targeting technicians, veterinarians, and laboratory staff, contacted Arluke about publishing the talk he'd delivered at the meeting. "The editors said that they would not change the content, but the title 'Stress Among Lab Researchers' was too controversial, and could I change it?"

Arluke, bemused by the uproar, asked the editors for suggestions, noting that he was "running out of synonyms." The editors of *Lab Animal* suggested *uneasiness,* and the article was published with that title. After the article appeared, Arluke was invited to talk about his work at a major pharmaceutical company. "I was asked how I would like to title my talk. I suggested 'Uneasiness Among Animal Researchers,' since that worked for *Lab Animal*. The people who invited me said, 'That would be too controversial; it would inflame research directors if you said that anyone was uneasy here.'" Arluke's hosts suggested that he call the talk "How Researchers Deal with Feelings," and he compiled, noting, "After this, I decided to leave all future talks untitled."

Arluke believes that his experience is "symptomatic of a taboo in science about talking about or acknowledging the complexity of human-animal relationships in laboratories." Nonetheless, he is a popular lecturer at research institutions, suggesting that even if companies and universities are not yet willing to copy the Spinelli model to provide their employees with the opportunity to discuss these issues, they are at least willing to entertain the possibility that their employees may have feelings about their work and would benefit from hearing about how others cope with their emotional response to the work and to public perceptions.

This seems a peculiarly American phenomenon. It does not fall under the category of "policy," yet these discussions may have as significant an impact on the actual well-being of laboratory animals as laws or regulations promulgated by Congress. USDA inspectors may visit a laboratory once a year, but it is the attitudes and behavior of the investigators, veterinarians, and technicians who staff the lab that determine the quality of care received by the animals who live and die there. For example, rats and mice, which comprise nearly 90 percent of the animals used in U.S. research laboratories, remain unprotected by the Animal Welfare Act. Animal protection groups have lobbied for years to include rats, mice, and birds under the regulations of the Act, against the strenuous objections of the USDA, which argues that it simply does not have the staff or resources to take on the responsibility of monitoring the husbandry and care of millions of rodents. Nonetheless, early in 1999 a request appeared on COMPMED, an Internet mailing list for veterinarians and laboratory personnel that has thousands of subscribers, for criteria to identify the presence of pain or distress in rats and mice. The question indicates a growing awareness that laboratory rats and mice are capable of suffering and that their suffering is morally significant and ought to be alleviated. As Joseph Spinelli has noted, "to prevent suffering in laboratory animals, we must first believe that they can suffer. Then we must dedicate ourselves to reducing or eliminating situations in which they suffer."

The same might be said of laboratory workers themselves,

who are often characterized by those who disapprove of their work as unfeeling automatons or outright sadists. If researchers could bring themselves to discuss their feelings about their work rather than simply reciting a litany of important discoveries based on animal research, they might find greater support for that work as the public begins to recognize that it is carried out by human beings who experience a sometimes painful mix of emotions and not by supermen and superwomen somehow immune from ordinary human fears, discomfort, and insecurities.

Edward Walsh, the researcher who endured years of harassment by animal rights campaigners objecting to his research, can testify to the difficulties of that endeavor. "Many friends and colleagues have urged me not to confess my feelings," he said, "not to admit that the threats and intimidating actions of PETA and its acolytes produce fear and uncertainty. Somehow, they conclude that an honest disclosure of the impact that the animal rights movement has had on our family will either raise the specter of personal danger to a higher level or convey an image of weakness, of vulnerability, ultimately inspiring more aggression and escalating the level of their terrorism. To them, I say no. I reveal my frailty to demonstrate my resolve. Prevarication and deception cannot be rewarded in a secure society."

Admitting that "experimenting and sacrificing animals for science has always been difficult for me," Walsh said that "those who believe that scientists casually conscript animals into science have either been misled or have actively rejected the truth" and that "wrestling with the cost versus the benefit of using an animal in an experiment is a way of life for all compassionate scientists. Our practice is and always has been to develop a plan that minimizes the number of animals required to complete a study and uses the least invasive, least traumatic procedures possible to answer the question. Over the course of our careers, JoAnn and I have progressively sought procedures and policies that improve the care and treatment of our animals because it is the only compassionate, responsible attitude. We will continue that practice into the future."

Because of the harassment experienced by researchers like Ed Walsh and JoAnn McGee, laboratory workers often hesitate to speak publicly about their work, even among friends and family. "Most people want to keep a low profile rather than go out and fight the good battle" to defend their work, Arnold Arluke has found. "Hearing about laboratory break-ins and death threats to workers around the country understandably creates a major chilling effect in many laboratories. Thus, people are reluctant to speak publicly because of possible danger to themselves, their workers, or their laboratories. Many scientists and technicians also said that they felt uncomfortable taking a stance in public because it was not the role of scientists to do so. And even if they tried, many felt ill equipped to speak at the local town hall on this issue. Others felt that speaking in public was a losing battle because the emotional arguments of those opposed to animal research would always trump their own more rational defenses."

Despite these dangers, an effort by scientists and laboratory workers to speak openly about animal research and to explain why they continue the work in light of their own sometimes conflicting emotions might help defuse much of the hostility and misunderstanding surrounding animal experimentation. Laboratory workers are frontline soldiers in the ethical battle that confronts anyone who seriously contemplates the issues raised by animal protectionists over the past hundred and fifty years. We love animals—and we eat them. We recognize our kinship with them and their ability to experience pain and pleasure—and we experiment on them. How do we balance our own needs and right to protect ourselves from hunger, disease, and death with the needs and rights of animals to be "subjects of a life," as animal rights philosopher Tom Regan has phrased it?

Those who have considered this question deeply, and grappled with the profound moral dilemma that lies at the heart of the question, have formulated different responses, as they have to the many other ethical conundrums confronting human beings. The very fact that we are able to entertain these questions at all points to a key difference between human beings and other animals.

Whether that difference is profound enough to enable us to continue to use animals as food, as experimental subjects, as the means to our ends, is something that each of us must contemplate and decide in the private space of individual conscience—recognizing that our answer to that question may well conflict with those of other people of good will and conscience. This ambiguous and unsettling outcome may be the best that we can hope for until our unquenchable human ingenuity finds a path beyond the conflict and a solution to which all can, in good conscience, agree.

EPILOGUE

Mr. Goldstein, who has been charged with second-degree mur-
der in the death of Ms. Webdale, has spent the last decade in
and out of state, city and private hospitals, as well as halfway houses
and outpatient clinics for treatment of mental illness, law
enforcement and health officials said. . . . Police officials said
that Mr. Goldstein told detectives that he uses two kinds of med-
ication, Haldol and Congentin, for his mental illness. Officials
would not comment on the specifics of Mr. Goldstein's illness,
but the two drugs are often used to treat paranoid schizophrenia.
—Michael Cooper,
The New York Times, January 6, 1999

CATONSVILLE, MARYLAND, 1999

ABOUT A MONTH after Andrew Goldstein pushed Kendra
Webdale in front of a subway train in New York City, I visited the Maryland Psychiatric Research Center (MPRC), where
researchers working in a variety of disciplines study the disease
that tormented Goldstein and led to Webdale's death. William T.
Carpenter, director of the MPRC and an internationally recognized expert on the disease, calls schizophrenia "the worst disease afflicting mankind, including AIDS." There are more than
two million schizophrenics in America alone, and the costs of
lifetime treatment and lost productivity are staggering, approximately $32.5 billion a year. The human costs, never mind the
economics, are devastating. "Schizophrenia destroys what is
distinctly human in a highly stigmatizing way," says Carpenter.

Symptoms usually appear between the ages of fifteen and twenty-five, with delusions, hallucinations, and disturbances in emotional and social functioning. The causes of schizophrenia remain unknown. Although research has shown that the disease has a strong genetic component, environmental insults (for example, a prenatal virus or developmental abnormalities) may play a role in triggering it as well.

During a visit to the MPRC, I held a human brain in my hands. It was an old brain, preserved long enough to have become beige and thick and rubbery. Fresh brains have a more gelatinous texture, the neuroscientist in charge of the laboratory told me, as I turned the brain over in my gloved hands. She pointed out the important parts—the forebrain, where higher intellectual functioning is carried out; the hippocampus, which plays an important role in memory; the thalamus and hypothalamus buried deep in the center of the organ, where they relay sensory stimuli and regulate body functions. As we talked and I turned the three-pound brain this way and that, I couldn't help but think that despite all the attention and acclaim the brain receives, this much-revered bit of human anatomy is really just a hunk of meat.

It reminded me of something a medical student had recently told me, describing freshman anatomy, the introductory course in which each doctor-to-be is assigned a cadaver. This particular student had never seen a dead body before, so the very sight of his first corpse was disturbing, he said. Over the next few months he found himself experiencing a complex series of emotions as he slowly explored the nooks and crannies of the human body. "First, you're shocked," he said, "then sad, then bored, then shocked again when you realize how accustomed you've become to the whole process, when you lift your head one day and realize that you've spent a whole afternoon poking around in someone's chest cavity."

This student had studied intellectual history as an undergraduate, and he now spent a lot of time pondering the emotional and philosophical ramifications of his work. He was concerned about the dehumanization that he felt was an inevitable outcome of learning to see people as simply bodies, and the "deromanticiza-tion" of life that he saw as its effect. At the same time, he felt that it was a great honor and privilege to be permitted to carry out this endeavor, and he felt grateful to the people who had donated their bodies to science so that future doctors might learn from them. That altruistic urge to facilitate research and learning was somehow also embedded in the tissue I held in my hands; the brain had belonged to a schizophrenic whose family donated the organ to the Maryland Brain Collection, an arm of the MPRC. Neuroscientists there are conducting ultrastructural studies of the tissue of the six hundred brains in the collection, peering deep inside brain cells, searching for pathological changes that might provide clues to the causes of schizophrenia, or to its more effective treatment.

Despite the admittedly macabre aspects of standing in a roomful of giant silver refrigerators stuffed with human brains, I felt optimistic there, knowing that researchers are working hard to try to understand this dreadful disease, which destroys so many lives. Like most illnesses, schizophrenia affects not just the lives of patients, but also those of family and friends who watch their loved ones lose competency and individuality. In many cases schizophrenics become paranoid and violent, like Andrew Goldstein, and are tortured by taunting voices and obsessive thoughts that urge them to harm themselves or others. Some become apathetic and unreachable, profoundly isolated from the world around them. Many of the drugs that are currently used to treat schizophrenia have unpleasant side effects, creating the abnormal muscle movements and tremors known as tardive dysk-inesia. The key to understanding schizophrenia, which appears throughout the world in every culture in roughly the same pro-portion, 1 percent of the population, is locked inside the kind of tissue I held in my hands, in the brains of dead schizophrenics—and of those living patients who consent to participate in re-search studies. Animals, too, are used in this and other types of neuroscience research, including studies of addiction, learn-ing disorders, stroke, trauma, Alzheimer's disease, and epilepsy.

According to the Society for Neuroscience, more than fifty million Americans suffer from a permanent neurological disability, and more people are hospitalized as a result of neuropsychiatric disorders than any other disease category.

Research on schizophrenia and other neurological disorders presents a moral dilemma having little to do with animals. Neuroscientists at MPRC and other research institutions have been targeted by activists who object to the use of mentally ill people as research subjects, even when patients are willing to participate in such studies. Even those who do not object in principle to such research are troubled by the lack of special protection for especially vulnerable populations like the mentally ill. Although the Belmont Report, formulated by the National Commission for the Protection of Human Subjects of Biomedical and Behavioral Research in 1974, laid down a general set of ethical principles and guidelines for the protection of human subjects, the United States still "has no specific policy on research involving persons with mental disabilities, a population that includes individuals with mental illness, developmental disabilities, dementia, and other conditions associated with mental impairment," says Rebecca Dresser, an attorney and bioethicist who has written extensively on these subjects.

The responsibility for monitoring research on these vulnerable subjects falls on institutional review boards, which "lack adequate guidance on fulfilling this responsibility," Dresser says. But as in animal research, simple either-or solutions fail to address the complexity of the issues at stake. "Though current protection may be inadequate," says Dresser, "a policy prohibiting research on mentally disabled subjects would create serious ethical problems of its own. If subjects with psychiatric disorders, dementia, and other mental impairments were excluded from studies, persons with these conditions would be deprived of the improved health care research can produce. Thus, the ethical challenge is to devise polices that permit promising research to continue, while protecting the rights and welfare of persons participating in such studies."

Washout studies, in which patients are taken off all medication for a period to see if symptoms recur, or to clear their systems before new drugs or drug combinations can be tried, have been the focus of particular concern. The questions raised by this type of research are not easy to resolve, but then again, the ethical and philosophical conundrums posed by biomedical research have always been harrowing. Human beings, like animals, have served as experimental subjects since the dawn of scientific medicine. The laws protecting both humans and animals are more stringent today than they have ever been, and investigators must explain and justify their experimental protocols (what they are doing, to whom, and why) before critical committees of their peers. This process is meant to protect not only the research subject but also the institution and the investigator. It is not a perfect system, but it does offer an opportunity to balance research needs with the rights of the subject. Difficult questions remain.

What should be permitted and what should be forbidden? How far can we extend the rights of animal subjects without seriously compromising the ability to conduct research at all? How valid is the informed consent of a human being who is ill and perhaps desperate for any improvement? Finally, and more generally, are we simply the depersonalized matter that biomedicine treats and studies, hog-tied by our DNA, and prey to any nasty virus that chooses to burrow into our cells? Or are we somehow more than the sum of our sometimes diseased parts? Art, literature, philosophy, and ethics—human activities that seek to provide answers to these questions—seem like hazy, insubstantial entities when you are holding a human brain in your hands. Yet all these marvels, and science itself, were produced by this three-pound mass, which looks so very unassuming. "There's a divinity that shapes our ends, Rough-hew them how we will," said Hamlet, in a prescientific age. Increasingly, those ends appear to be embedded in our cells, and our bodies, these messy vulnerable chunks of matter pulsating with energy and biochemical activity, prime movers of our destiny. This realization disturbs many people, and always has.

"Generally, we refuse to admit within ourselves, or within our friends, the fullness of that pushing, self-protective, malodorous, carnivorous, lecherous fever which is the very nature of the organic cell," writes Joseph Campbell in *The Hero with a Thousand Faces*, his great study of the human monomyth, the quest for enlightenment found in varying forms in every culture. "When it suddenly dawns on us or is forced to our attention, that everything we think or do is necessarily tainted with the odor of the flesh, then, not uncommonly, there is experienced a moment of revulsion: life, the acts of life, the organs of life, woman in particular as the great symbol of life, become intolerable to the pure, the pure, pure soul."

This stage of repulsion and disgust is often followed by a period of identification with some higher good, Campbell says, to which all the hero's strivings for purity and transcendence are attached. Subsequently, he denies and represses his knowledge of the messy reality of organic life, the shattering realization that humans, too, are "meat," teeming with microorganisms, some of them lethal, that all life feeds on life and that all life eventually ends in death, for the virtuous as well as the corrupt. Everything that the hero rejects as a result of his desire to transcend the pleasures and limitations of biological life, every unbearable negative, is then projected onto various others, who embody the despised traits that the hero works so hard to overcome, imagining that by doing so he will be immune to the common destiny. The stage is set for conflict.

"The battlefield is symbolic of the field of life, where every creature lives on the death of another. A realization of the inevitable guilt of life may so sicken the heart that, like Hamlet or Arjuna, one may refuse to go on with it. On the other hand, like most of the rest of us, one may invent a false, finally unjustified, image of oneself as an exceptional phenomenon in the world, not guilty as others are, but justified in one's inevitable sinning because one represents the good," Campbell writes. "Such self-righteousness leads to a misunderstanding, not only of oneself but of the nature of both man and the cosmos. The goal of the myth is to dispel the need for such life ignorance by effecting a reconciliation of the individual consciousness with the universal will. And this is effected through a realization of the true relationship of the passing phenomena of time to the imperishable life that lives and dies in all."

Individual lives, human or animal, inevitably end in death, but life itself continues in a endless spiral of birth, growth, death, and rebirth. "Nothing retains its own form; but Nature, the great renewer, ever makes up forms from forms. Be sure there's nothing that perishes in the whole universe; it does but vary and renew its form," Ovid wrote in one of the classics of Western literature, *Metamorphoses*. In an irreligious age, which both worships matter and despises it, this concept may seem too mystical to some. But it is the lesson taught by nature, the nature that science studies and the animal protection movement claims to champion. All are part of some greater whole, which some term Gaia, the living earth. Scientist or activist, animal or human, all return to her in the end. Perhaps it is she whom the protagonists of the struggle described in this book have been fighting all along, while they dreamed they were battling each other.

Science and animal protection have, each in its own way, created philosophical systems that challenge the limits of the natural and seek to impose a man-made order. Each has attempted to carry out a work of redemption, however differently they define the term. Inescapably embedded in nature, and at the same time standing outside the great round, analyzing and categorizing, human beings have sought throughout history to comprehend the universe and ourselves, to impose order on a world that can appear arbitrary and chaotic. Nature in her pure form is both terrifying and beautiful, nurturing and devouring. Her will is implacable and the fight, no matter how valiant, doomed to defeat. But in that paradoxical struggle (to fight and to submit, to impose order and to surrender to that which is beyond our control) the hero is born, according to Campbell: "Not the animal world, not the plant world, not the miracle of the spheres, but man himself is now the crucial mystery."

Epilogue

The quest to understand this mystery of human identity, to claim our proper place in the universe and our proper relation to other forms of life, remains the essential human task. Both scientists and animal protectionists have struggled with this riddle. The time has come for the rest of us to take up the challenge. "Every one of us shares the supreme ordeal," says Campbell. Each of us must somehow seek our own answer to our puzzling supra-animality, and this sublime enigma of human life.

NOTES

INTRODUCTION

3–4 *"the real question to ask"* Ron Kaufman, "Scientists Doubtful About New Law Aiming to Protect Animal Research Facilities." *The Scientist,* October 26, 1992, 1.

5 *"Many and long were the conversations"* Mary Godwin Shelley, introduction to *Frankenstein, or The Modern Prometheus* (Oxford: Oxford University Press, 1969), Ibid., 8.

5 *"I busied myself to think of a story"* Ibid., 8.

6 *"I saw—with shut eyes, but acute mental vision"* Ibid., 9.

6 *"To examine the causes of life"* Ibid., 51.

7 *In an era when "resurrection men" were removing fresh corpses from graves* See Ruth Richardson, *Death, Dissection, and the Destitute* (London: Routledge & Kegan Paul, 1987).

7 *"In a culture in which organ transplants, life-extension machinery"* Susan Bordo, *Unbearable Weight: Feminism, Western Culture and the Body* (Berkeley: University of California Press 1993), 245.

10 *"It was as if all the pain in the world had found a voice"* H. G. Wells, *The Island of Dr. Moreau* (New York: Penguin, 1987), 36.

12 *"the writing of* Higher Superstition *was undertaken"* Paul R. Gross and Norman Levitt, *Higher Superstition: The Academic Left and Its Quarrels With Science* (Baltimore and London: Johns Hopkins University Press, 1994), ix.

15 *"When Shelley pictured science as a modern Prometheus"* Jacob Bronowski, *Science and Human Values* (New York: Harper & Row, 1965), 5.

1: VIRUSES, VACCINES, AND VIVISECTION

17 *"To learn how men and animals live"* Claude Bernard, *Introduction à l'étude de la médecine expérimentale (An Introduction to the Study of Experimental Medicine)* (New York: Macmillan, 1927), 99.

17 *How can experiments on dogs and sheep and rabbits* See Bernard's defense of vivisection in the above text, 99–103.

18 *"The experiment itself supplies me with the key questions"* Ibid., 151–162.

19 *The men who locked themselves in basement and attic laboratories* See Paul Elliott, "Vivisection in Nineteenth Century France," in *Vivisection in Historical Perspective*, Nicolaas Rupke, ed. (London: Routledge, 1987), 52.

20 *an experimenter did not endanger his own soul* See Andreas-Holger Maehle and Ulrich Trohler, "Animal Experimentation from Antiquity to the End of the Eighteenth Century: Attitudes and Arguments," in Rupke, 21.

20 *a machine that contains its own principles of motion* See Jean-Claude Beaune, "The Classical Age of Automata: An Impressionistic Study from the Sixteenth to Nineteenth Century," in *Fragments for a History of the Human Body, Part One*, Michael Feher with Ramona Nadoff and Nadia Tazi, eds. (New York: Urzone), 431.

20 *"The capacity of animals for sensation"* John Parascandola, "The History of Animal Use in the Life Sciences," in *Proceedings of the World Congress on Alternatives and Animal Use in the Life Sciences: Research, Education and Testing* (New York: Mary Ann Liebert, 1994), 13.

21 *to "get in contact with the great teachers"* Donald Fleming, *William Henry Welch and the Rise of Modern Medicine* (Baltimore: Johns Hopkins University Press, 1954), 75.

23 *although it was said that he was far more sensitive* See Rom Harre, *Great Scientific Experiments: 20 Experiments that Changed Our View of the World* (Oxford: Phaidon, 1981), 103. "In subsequent work on more virulent diseases such as rabies, he increasingly turned to the help of assistants, partly because of his revulsion from the necessary vivisection the research required."

25 *near riots had broken out* See Richard French, *Antivivisection and Medical Science in Victorian Society* (Princeton: Princeton University Press, 1975), 20–21.

25 *As early as 1831, the British physiologist Marshall Hall* Andrew Rowan, *Of Mice, Models and Men: A Critical Evaluation of Ani-*

mal Research (Albany: State University of New York Press, 1984), 45–46.

26 *"In that laboratory we sacrificed daily"* George Hoggan, *Morning Post*, February 2, 1875, reprinted as Appendix II in French, 414–415.

26 *The book made no mention of anesthesia* See Barbara Orlans, *In the Name of Science: Issues in Responsible Animal Experimentation* (Oxford: Oxford University Press, 1993), 16.

28 *and a similar statistical discrepancy between men and women* See Harold A. Herzog, "Sociology of the Animal Rights Movement," in *Encyclopedia of Animal Rights and Animal Welfare*, Marc Bekoff, ed. (Westport, Conn.: Greenwood Press, 1998), 53–54.

28 *"the abolition of bull-baiting interfered with the amusement of the people"* Charles D. Niven, *History of the Humane Movement* (London: Johnson, 1967), 58.

28 *"no reason can be assigned for the interference of legislation"* Ibid., 60.

28 *As early as 1827, the SPCA issued a call* See French, 28.

2: THE KINGDOM OF THE SPIRIT

30 *"I went in my sleep last night"* Anna Kingsford, diary entry, February 2, 1880, cited in Edward Maitland, *The Life and Times of Anna Kingsford: Her Diaries, Letters, and Work* (London: Watkins, 1896).

30 *But Anna knew that she had important work to do here* The depiction of Anna Kingsford in this chapter is based largely on Maitland's two-volume *Life*. Most of the events described in this chapter are based on accounts in that book. Unless otherwise noted, all quotes attributed to Kingsford are from her diary. Other sources include E. Westacott, *A Century of Vivisection and Antivivisection* (Essex, England: C. W. Daniel Co., 1949) and John Vyvyan, *In Pity and in Anger: A Study of the Use of Animals in Science* (London: Michael Joseph, 1969).

34 *Cobbe, an Irishwoman twenty-four years older than her editor* Information on the life and personality of Frances Powers Cobbe is largely based on her autobiography, *Life of Frances Powers Cobbe as Told by Herself* (London: Swan, Sonnenschein and Co., 1904), and her writing, including *The Modern Rack* (London: Swan, Sonnenschein and Co., 1889). The rivalry between Kingsford and Cobbe is also covered by Maitland, Westacott, and Vyvyan.

34 *Cobbe, although an ardent antivivisectionist* French, 230–31.

37 *"If the walls came tumbling down"* James Turner, *Reckoning with the Beast: Animals, Pain and Humanity in the Victorian Mind* (Baltimore: Johns Hopkins University Press, 1980), 61.

37 *"Suffering may be everywhere"* Judith Perkins, *The Suffering Self: Pain and Narrative Representation in the Early Christian Era* (London: Routledge, 1995), 7.

37 *"The dread of pain"* Turner, 79.

39 *"Except for teaching purposes"* Testimony of Emanuel Klein to the Royal Commission, reprinted in French, 103–6.

41 *"Miss Cobbe had painfully reconstructed a faith"* Turner, 90.

41 *"Round and fat as a Turkish sultana"* Ibid., 89.

42 *"We had been discussing Evolution"* Cobbe, quoted in John Vyvyan, *In Pity and in Anger: A Study of the Use of Animals in Science* (London: Michael Joseph, 1969), 72.

44 *"The years 1875 and 1876 saw the beginnings of the hysteria"* French, 82.

44 *"the society would watch the existing Act"* E. Westacott, *A Century of Vivisection and Antivivisection: A Study of Their Effect upon Science, Medicine and Human Life During the Past Hundred Years* (Essex, England: C. W. Daniel, 1949), 124.

44 *"accepted blindly the representation of vivisection"* Cobbe, in Westacott, 125.

45 *"vivisection we recognized at last to be a* method of research" Ibid., 125.

45 *"if they had been able to cooperate"* Vyvyan, 136–37.

48 *To Blavatsky she confided* Westacott, 180–181; Vyvyan, 138–40.

49 *"Let us work against the* principle *then"* Westacott, 181.

49 *"Although presaged by the activity of men like Dalton and Flint"* Turner, 92.

50 *Although Henry Bergh* Susan Lederer, *Subjected to Science: Human Experimentation in America Before the Second World War* (Baltimore: Johns Hopkins Press, 1995), 237.

50 *"Perhaps more significant, Caroline White's assistant editor"* Susan Lederer, "Controversy in America, 1880–1914," in *Vivisection in Historical Perspective*, Nicolaas Rupke, ed. (London: Routledge, 1987), 238–39.

50 *The New England Antivivisection Society was founded* Ibid., 238–39.

51 *women's feelings of powerlessness, resentment, and anger* See Coral Lansbury, *The Old Brown Dog: Women, Workers and Vivisection*

in Edwardian England (Madison: University of Wisconsin Press, 1985).

52 *"Women alone suffered the degradation of internal inspection"* Mary Ann Elston, "Women and Antivivisection in Victorian England, 1870–1900," in Rupke, 277.

52 *"appealed to the same kinds of fears"* French, 229.

52 *"all three [movements] were campaigns"* Elston, 274.

52 *"The death rates from ovariotomy"* Ibid., 278.

52 *"the metaphor of medical science, and medical practice on women"* Ibid., 279.

53 *"Much of the treatment prescribed by physicians"* Carroll Smith-Rosenberg, "The Hysterical Woman: Sex Roles and Role Conflict in Nineteenth Century America," in *The Yellow Wallpaper: Women Writers Text and Context Series,* Thomas L. Erskine and Connie L. Richards, eds. (New Brunswick, N.J.: Rutgers University Press, 1992), 93.

53 *"At first, and in some cases for four or five weeks"* S. Weir Mitchell, "Fat & Blood: An Essay on the Treatment of Neurasthenia and Hysteria" (Philadelphia: J. B. Lippincott, 1877), excerpts printed in *Women Writers: Text and Context Series,* 106.

54 *"The hysterical female emerges from the essentially male medical literature"* Smith-Rosenberg, 95.

3: THE DOGS OF WAR

55 *"It is a well-known fact"* Simon Flexner and James Thomas Flexner, *William Henry Welch and the Heroic Age of American Medicine* (New York: Viking Press, 1941), 362.

56 *But public health officers in most cities in the United States* See Elizabeth Fee, *Disease and Discovery* (Baltimore: Johns Hopkins University Press, 1987), 28.

57 *knowing that Popsy seldom prepared written notes* Flexner and Flexner, 235–36.

58 *"Whereas the dog has made a wonderful war record"* The testimony of Welch and other speakers at the hearing is reproduced verbatim from the hearing transcript, "Prohibiting Vivisection of Dogs," Senate Committee on the Judiciary, Bill 66 S. 1258, November 1, 1919.

61 *"As long as the noble nature and the selfish nature"* Flexner and Flexner, 50.

62 *"of the required three years of study"* Ibid., 62.

62 *Welch spent long days and nights in "the dead house"* Ibid., 70.

63 *"We have nothing in America like these laboratory courses"* Ibid., 80.

63 *"My work with Professor Ludwig has been very profitable"* Ibid., 85.

64 *"I proved first that it is possible to produce oedema"* Ibid., 96–97.

65 *"As no human power can change the inevitable"* Ibid., 88.

72 *"the most promising approach to the study of nutritional requirements"* J. R. Lindsey, "Historical Foundations," in *The Laboratory Rat*, vol. 1 (New York: Academic Press, 1979), 12.

75 *"The problems of infectious diseases in animals"* Flexner and Flexner, 184.

76 *"Mus musculus, the house mouse of North America and Europe"* Herbert C. Morse, "The Laboratory Mouse—An Historical Perspective," in *The Mouse in Biomedical Research*, vol. 1 (New York: Academic Press, 1981), 1.

77 *"served as the initial testing ground"* Lindsey, 6.

77 *"The 1920's saw the advent of the vast majority of inbred strains"* Morse, 10–11.

78 *"they were determined that he should demonstrate that alcoholism"* Ibid., 10.

78 *"by virtue of certain pioneering work"* Flexner and Flexner, 9.

4: NAZI HEALING

80 *"Beginning with general hygiene"* Deutsches Ärtzeblatt, June 1933, quoted in Hartmut M. Hanauske-Abel, "Not a Slippery Slope or a Sudden Subversion: German Medicine and National Socialism in 1933," *British Medical Journal* 313 (1996): 1453–63, 1996.

82 *Galton believed that if scientific principles were rationally applied to populations* See Daniel J. Kevles, *In the Name of Eugenics: Genetics and the Uses of Human Heredity* (New York: Alfred A. Knopf, 1985).

83 *Natural selection "proposes no perfecting principles"* Stephen Jay Gould, "Darwin's Untimely Burial," in *Ever Since Darwin: Reflections in Natural History"* (New York and London: W. W. Norton, 1977), 45.

84 *In the 1930s however, the progressive fallacy was widespread* See Daniel J. Kelves and Stefan Kuhl, *The Nazi Connection: Eugenics, American Racism, and German National Socialism* (Oxford: Oxford University Press, 1993).

84 *By 1939, the Nazi Hereditary Health Courts* Information in this chapter on Nazi euthanasia and sterilization programs is based on accounts in Robert J. Lifton, *The Nazi Doctors: Medical Killing and the Psychology of Genocide* (New York: Basic Books, 1986); James M. Glass, *Life Unworthy of Life: Racial Phobia and Mass Murder in Hitler's Germany* (New York: Basic Books, 1987); and Robert Proctor, *Racial Hygiene: Medicine Under the Nazis* (Cambridge: Harvard University Press, 1988). Another important source is a special issue of the *British Medical Journal* published in December 1996 to commemorate the fiftieth anniversary of the Nuremburg medical trials (volume 313, December 7, 1996), 1413–75.

85 *"There was wonderful material among these brains"* J. Hallervorden, quoted in Hanauske-Abel, 1457.

85 *"Euthanasia was there to help the sufferer"* Karl Brandt, quoted in Paul Hoedeman, *Hitler or Hippocrates: Medical Experiments and Euthanasia in the Third Reich*, translated from the Dutch by Ralph de Rijke (Sussex: The Book Guild, 1991), 115.

86 *"The use of the term 'Nazi' implies historical uniqueness"* Hanauske-Abel, 1453.

86 *"On the one hand, the Nazis wanted to return"* Proctor, 243.

87 *"Nazi encouragement for new and organic forms of medicine"* Proctor, 227. For an account of German holistic healing and its tensions with scientific medicine, see also Anne Harrington, *Reenchanted Science: Holism in German Culture from Wilhelm II to Hitler* (Princeton: Princeton University Press, 1996).

87 *The ban on vivisection* See Arnold Arluke and Clinton R. Sanders, *Regarding Animals* (Philadelphia: Temple University Press, 1996), and Luc Ferry, "Nazi Ecology," in *The New Ecological Order*, translated by Carol Volk (Chicago and London: University of Chicago Press, 1995).

87 *These laws "permitted experiments on animals in some circumstances"* Arluke and Sanders, 134.

88 *"helped to shape the Third Reich's criticisms"* Ibid., 143.

88 *"Hitler was a vegetarian and did not smoke or drink"* Proctor, 228.

88 *"many leading Nazis practiced vegetarianism"* Arluke and Sanders, 150.

88 *"Meat-eating is a perversion"* Joseph Goebbels, quoted in Arluke and Sanders, 150.

89 *"people were constantly twitting me about why Hitler was not included"* This and quotes following: Rynn Berry, conversation with the author, November 1999. See also his pamphlet "Why

Hitler Was Not a Vegetarian" (New York and Los Angeles: Pythagorean Publishers, 1999).

91 *"Out of respect for human life"* Fritz Klein, quoted in Hoedeman, 193.

91 *A range of experiments were conducted at various camps* For an official account of human experiments under the Nazi regime, see *Trials of Major War Criminals before the Nuremburg Military Tribunals,* fifteen volumes (Washington, D.C.: U.S. Government Printing Office, 1949–1952). For summaries, see Lifton and Hoedeman.

92 *"If one desires to judge a particular behavior"* Fritz Ernest Fischer, quoted in Hoedeman, 72.

92 *"It was not cultural propagandists who organized the infamous 'special treatment' of the Jews"* Glass, 7.

93 *"It was not that the Germans were unaware of typhus as a disease process"* Ibid., xvi.

93 *"The Final Solution, from the perpetrators' perspective"* Ibid., 49.

93 *"The book included as 'unworthy life' "* Lifton, 46.

93–94 *"One can speak of the Nazi state as a 'biocracy' "* Ibid., 32.

94 *"to care about public health"* Glass, 186.

95 *"What distinguishes the Final Solution"* Ibid., 60.

96 *Three hundred and ninety-nine men in Macon County, Alabama* See James H. Jones, *Bad Blood: The Tuskegee Syphilis Experiment* (New York: The Free Press, 1981).

5: POLIO POLITICS

98 *"Animal experimentation in our country is assuming"* Letter from Robert Gesell to A. J. Carlson, from the archives of the Animal Welfare Institute.

98 *Christine Stevens and her father, Dr. Robert Gesell, prepared to enter the research laboratories* Information on Robert Gesell and the founding of the Animal Welfare Institute is based on the author's conversations with Christine Stevens in the summer of 1997 and on the archives (including private correspondence and newsletters) of the Animal Welfare Institute.

99 *"I am not one of those who believe that the conditions"* Letter from Gesell to Carlson, February 8, 1946. The National Society for Medical Research was founded in 1945 by Dr. A. J. Carlson, Dr. Andrew C. Ivy, and Dr. George Wakerlin.

100 *"These ominous experiments make us search our souls"* Letter from Gesell to Carlson, August 1949.

100 *"On the urgent problem of laboratory animal"* Ibid.

101 *"That scientists should not be hampered and animals completely protected"* AWI Information Report, June–July 1952.

102 *"What we need is people to take care of the animals"* AWI Information Report, March–April 1954.

102 *The manual for the care of laboratory animals* This book, *Basic Care for Experimental Animals*, was published by AWI in 1953.

103 *In response to Salk's inquiry two years before.* The narration of the founding of the National Foundation for Infantile Paralysis and the search for a polio vaccine is based on the accounts provided by Jane S. Smith in *Patenting the Sun: Polio and the Salk Vaccine* (New York: William Morrow, 1990), Aaron E. Klein in *Trial by Fury: The Polio Vaccine Controversy* (New York: Charles Scribner's Sons, 1972), and J. R. Paul in *History of Poliomyelitis* (New Haven, Conn., and London: Yale University Press, 1971).

105 *"The initial impact which the NFIP made upon that minuscule number"* Paul, 311.

105 *"Between 1938 and 1962, the Foundation's annual income"* Ibid., 312.

105 *Six years after its founding* Ibid., 319.

106 *"One of the achievements brought about by the effective use of propaganda techniques"* Ibid., 311–12.

106 *"The corner drug store may have a good supply"* Pittsburgh Sun-Telegraph, December 12, 1954; reprinted in AWI Information Report, January–February 1955.

107 *"A shipment of 900 rhesus monkeys"* AWI Information Report, January–February 1955.

108 *"in addition to the cruelty"* Ibid.

109 *"The practice of Machiavelli's rule of deliberate hypocrisy"* Ibid.

109 *"to make the utterly ridiculous claim that scientists somehow are exempt from the cruelty-to-animals laws"* In fact, although most states passed anticruelty statutes between 1828 and 1898, fourteen of these laws specifically exempted animal experiments. The only two instances in which researchers were successfully prosecuted under anticruelty statutes occurred in Massachusetts in 1958 and Maryland in 1981. The latter conviction was overturned on the grounds that federally funded biomedical research was not covered by the law. See Christine Stevens, chapter 4, "Laboratory Animal Welfare," in *Animals and Their Legal Rights*, fourth edition (Washington, D.C.: Animal Welfare Information Center, 1990), 66.

109 *"Everyone hopes to see an end to polio"* AWI Information Report, January–February 1955.

110 *"between 1953 and 1960, well over a million monkeys died"* Andrew N. Rowan, *Of Mice, Models and Men: A Critical Evaluation of Animal Research* (Albany: State University of New York Press, 1984), 110.

111 *"9,000 monkeys, 150 chimpanzees"* Albert B. Sabin, "Present Status of Attenuated Live-Virus Poliomyelitis Vaccine," *Journal of the American Medical Association*, vol. 62, no. 18 (1956): 1589–96.

112 *"For a generation, starting with Flexner's early experiments"* Paul, 382.

112 *"Put simply, paralytic polio was an inadvertent by-product of modern sanitary conditions"* Smith, 35.

113 *"Before, polio researchers had to infect individual monkeys"* Smith, 127.

114 *"I look upon it as ritual and symbolic"* Ibid., 136.

115 *The scientific report announcing the vaccine's safety and efficacy* See Klein, 109.

116 *Investigations by the Communicable Disease Center* On the heels of the Cutter incident, the Communicable Disease Center was renamed the Centers for Disease Control and Prevention. The CDC is responsible for monitoring outbreaks of communicable diseases. In recent years, the outbreak of hantavirus in the Four Corners area of the American Southwest and the first cases of pneumocystis pneumonia and a rare skin cancer in gay men (early manifestations of the AIDS epidemic) were reported in CDC publications.

116 *Of that number, 79 children developed polio* See Smith, 266.

116–117 *"When compared to the 5,000,000 people who had received the vaccine"* Klein, 119.

117 *"Paralytic poliomyelitis is a phenomenon of the twentieth century"* Smith, 43.

118 *This comic book defender of justice* Superman was introduced to the comic book audience in 1938, Batman in 1939. Superman's dominance of the market continued until the mid-1960s. See *Seduction of the Innocent* by Frederic Wertham (New York: Rinehart, 1954); *Batman: The Complete History* by Les Daniels and Chip Kidd (San Francisco: Chronicle Books, 1999), and *Superman: The Complete History* by Les Daniels and Chip Kidd (San Francisco, Chronicle Books, 1998).

SIX: THE NEW CRUSADERS

120 *"Animal liberation is also human liberation"* Henry Spira, "Fighting for Animal Rights: Issues and Strategies," in Harlan B. Miller and William H. Williams, *Ethics and Animals* (Clifton, N.J.: Humana Press, 1983), 373.

120 *Henry Spira was all fired up* The account of Henry Spira's life and campaigns in this chapter is based on the author's conversations with Henry Spira, conducted from 1992 to 1998, and on his published work, much of it reprinted in his *Strategies for Activists: From the Campaign Files of Henry Spira* (New York: Animal Rights International, 1997). Spira died in September 1998.

122 *"What can be done within a job trust . . . ?"* Henry Spira, "Rule and Ruin in the NMU," *The Call for Union Democracy*, June 1969; reprinted in Spira's *Strategies for Activists*.

122 *heard the philosopher expound concepts from his manuscript in progress* The manuscript became the book *Animal Liberation* (New York: Avon Books, 1975), which is often referred to as the bible of the animal rights movement.

123 *Spira had his insider* Although Henry Spira often referred to Leonard Rack as one of the most brilliant people he had ever met, others did not share that opinion. Alan M. Goldberg, director of the Johns Hopkins Center for Alternatives to Animal Testing, says that Rack, who committed suicide prior to Spira's Revlon campaign, was a troubled man who was nothing like the scientific insider Spira imagined him to be.

124 *"Japan's use of laboratory animals"* Rowan, 64–65.

124 *"By 1983 over 80 Nobel laureates"* Victoria A. Hardin, *Inventing the NIH: Federal Biomedical Research Policy, 1887–1937* (Baltimore: Johns Hopkins University Press, 1986), 183.

125 *animal defenders like Brigid Brophy and Air Chief Marshal Lord Dowding and his wife, Muriel* See Andrew Linzey, "Brigid Brophy," in Marc Bekoff, ed., *Encyclopedia of Animal Rights and Animal Welfare* (Westport, Conn.: Greenwood Press, 1998), 96.

125 *"I firmly believe that painful experiments on animals"* Lord Dowding, quoted in Muriel Dowding, *The Psychic Life of Muriel, Lady Dowding: An Autobiography* (Wheaton, Ill.: The Theosophical Publishing House, 1980), 151.

126 *Her visits to U.S. laboratories still revealed dreadful conditions* See Christine Stevens, Chapter 4, "Laboratory Animal Welfare," in *Animals and Their Legal Rights,* fourth ed. (Washington: Animal Welfare Institute, 1990), 69.

126 *Ruth and Cox produced a 210-page report* See David N. Ruth and Robert W. Cox, *The Care and Management of Laboratory Animals Used in the Programs of the Department of Health, Education and Welfare* (Washington: Division of Operations Analysis, Office of the Comptroller, Office of the Secretary of Health, Education and Welfare, January 1966).

126–127 *Soon after, testimony in the United States Congress* See Stevens, 73.

129 *Despite long study and over $400,000 in federal grants* See Nicholas Wade, "Animal Rights: NIH Cat Study Brings Grief to NY Museum" *Science*, October 8, 1976, 134–37.

130 *"On the Fifth Floor of the Museum's Education Building"* Spira, *Strategies for Activists*, 129.

131 *"While I am not prepared at this moment"* Wade, 137.

131 *"We used a variety of tactics"* Spira, 129.

131 *"The museum's plight carries a warning"* Wade, 137.

132–133 *"We may well be entering the decade of animal rights,"* Spira, *Strategies for Activists*, 140.

133 *He wanted them to see the conditions in the laboratory* For Pacheco's version of events, see "The Silver Spring Monkeys," cowritten with Anna Francione, in *In Defense of Animals*, Peter Singer, ed. (London: Basil Blackwell, 1985).

134 *"Such a decision by NIH was not made lightly"* Barbara Orlans, *In the Name of Science: Issues in Responsible Animal Experimentation* (New York and Oxford: Oxford University Press, 1993), 178.

134 *Taub had been using a procedure* Orlans, 177–178. See also Deborah Blum, *The Monkey Wars* (Oxford: Oxford University Press, 1994), 107–8.

135 *Edward Taub's conviction sent shock waves through the research community.* In 1997, a decade and a half after his conviction, Taub was honored by the American Physiological Society for fundamental discoveries about brain organization. "Taub's award was based on research that he says is no longer permitted anywhere: he severed nerves in monkeys' arms to see what happens in the corresponding regions of their brains. . . . Taub and others have published research on the monkeys showing that their brains surprisingly underwent massive reorganization after their injuries. Based on this research, Taub has designed a routine to help people with disabilities from brain injuries that entails extensive exercise of the afflicted limb," "A Brighter Day for Edward Taub," in *Science*, vol. 276 (1997): 1503.

135 *According to a book written by PETA cofounder Ingrid Newkirk*
This book, *Free the Animals: The Inside Story of the Animal Liberation Front*, was published in 1987 by Noble Press. *People* magazine published an interview with a woman purporting to be "Valerie," the founder of the ALF, on January 8, 1993.

136 *"Valerie is totally fictitious"* Interview with Jo Shoesmith, July 2, 1998, Frederick, Maryland.

140 *"it added 100,000 new members each year"* James M. Jasper and Dorothy M. Nelkin, *The Animal Rights Crusade: Growth of a Moral Protest* (New York: The Free Press, 1992), 38.

140 *"In the first few years of the 1980's"* Susan Finsen and Lawrence Finsen, "The Animal Rights Movement," in Bekoff, 53.

140–141 *"there are now more than 400 animal advocacy groups"* Blum, 113–14.

141 *"The animal rights view"* Tom Regan, "Animal Rights," in Bekoff, 42–43.

142 *"Animal rights is a moral crusade"* Jasper and Nelkin, 7–8.

142 *"although far less numerous than pragmatist or welfarist organizations"* Ibid., 9.

142 *"By the early 1990's"* Blum, 113.

143 *"It seems to me that when a corporation is responsive to our concerns"* Spira, in Peter Singer, *Ethics into Action: Henry Spira and the Animal Rights Movement* (Oxford: Rowman and Littlefield, 1998), 131.

143 *"Despite its success, or perhaps because of it"* Jasper and Nelkin, 31.

143 *In a 1985 interview conducted by Charles W. Griswold, Jr.* "Ingrid Newkirk: Q and A," *Washington City Paper*, December 20, 1985. All Newkirk's quotes on the following pages are reprinted from the article.

147 *Over the next few years, the ALF continued raiding* The City of Hope Medical Center in Duarte, California, was fined $11,000 for violations of the Animal Welfare Act following an ALF raid in 1984. NIH suspended funding of the facility until problems were corrected. More than one hundred animals were stolen from the lab during the raid.

147 *John Orem, the sleep studies researcher* K. Mangan, "Universities Beef Up Security at Laboratories to Protect Researchers Threatened by Animal Rights Activists," *Chronicle of Higher Education,* September 19, 1990, 16–19.

148 *"The ALF was made up of very ordinary people"* Jo Shoesmith, interview with the author, July 1998.

7: STALKING THE SHADOW

150 *"While some traits peculiar to the shadow"* Carl Gustav Jung, *Aion: Researches into the Phenomenology of the Self* (Princeton: Princeton University Press, 1969), 9.

150 *A handsome, dark-haired man* This account of the arrest of Fran Stephanie Trutt for the attempted murder of Leon Hirsch is based on investigative reporting published in Connecticut newspapers (including the *Hour*, the *Advocate* and the *Westport News*). A full list of sources and reporter's names are listed in the References. These journalists not only interviewed the principals but also obtained transcripts of telephone conversations and copies of reports by undercover agents for Perceptions International. The general outlines of the story were confirmed in an article published in *The New York Times*, "Animal Rights Case: Terror or Entrapment?" published on Friday, March 3, 1989, by reporter Celeste Bohlen. The *Times* article also detailed the attempts of Perceptions International agent Mary Lou Sapone to incite other activists to violence in the months before her meeting with Fran Trutt.

158 *"it nonetheless appears"* Andrew N. Rowan, Franklin M. Loew, and Joan C. Weer, *The Animal Research Controversy: Protest, Process And Public Policy* (Boston: Tufts University School of Veterinary Medicine, 1995), 17.

158 *"Hoffman-LaRoche reported that its use"* Ibid., 18.

159 *In a 1998 interview with the author* Mark Matfield, interview with author, London, July 1998.

159 *After the passage of the Food, Drug and Cosmetic Act* See Ruth deForest Lamb, *American Chamber of Horrors: The Truth About Food and Drugs* (New York: Farrar and Rinehart, 1926).

162 *D. H. Henderson, a physician who had headed* This story is courtesy of Andrew Rowan, who was serving as science adviser to Henry Spira at the time.

164 *As late as 1997, representatives of the Environmental Defense Fund* See David Roe, *Toxic Ignorance* (Washington, D.C.: Environmental Defense Fund, 1997).

167 *"Our use of these terms"* W. M. S. Russell and Rex Burch, *The Principles of Humane Experimental Technique* (London: Metheun, 1959). Reprinted by the Universities Federation for Animal Welfare, 1992, 14.

167 *"Replacement means the substitution for conscious living higher animals"* Ibid., 64.

168 *"By now it is widely recognized"* Ibid., 5–6.

168 *"Russell and Burch's book had little impact"* Statement by Andrew Rowan in conversation with the author, March 1999.

169 *"We lost a lot of time debating the word alternatives"* Deborah Rudacille, "NIH Submits 3Rs Report to Congress," *Newsletter of the Johns Hopkins Center for Alternatives to Animal Testing,* vol. 12, nos. 1, 3.

169 *Despite this purported embrace* Herbert Lansdell, "The Three Rs: A Restrictive and Refutable Rigamarole," *Ethics and Behavior,* vol. 3, no. 2 (1993).

171 *"The research advocacy groups"* Rowan, Loew, and Weer, Andrew N. Rowan, Franklin Loew, and Joan Weer, *The Animal Rights Controversy: Protest, Process and Public Policy* (Boston: Tufts University School of Veterinary Medicine, 1995), 131.

171 *Surveys conducted by the National Science Board* Ibid., 12.

173 *"In light of this and other recent examples"* Health, Safety, and Research Alliance of New York State, *Alliance Update* (flier).

174 *"if you find it necessary to unfairly characterize"* Baldwin's quote originally appeared in the *Sunday Telegraph,* May 12, 1996, and was posted on the Americans for Medical Progress Web site.

175 *"It is not enough for us to wear red ribbons"* Joseph Murray, M.D., "Animals Hold the Key to Saving Human Lives," *Los Angeles Times,* February 5, 1996.

175 *"No AIDS breakthroughs have come out of animal research"* Sarah Ferguson "Glam Activism," *New York,* November 7, 1994, 64.

178 *public attitudes to the animal rights movement are mixed* Harold A. Herzog, Jr., "Sociology of the Animal Rights Movement," in Bekoff, 54.

8: SAINTS AND SINNERS

179 *"Many in the movement endorse 'diversity' "* Gary L. Francione, *Rain Without Thunder: The Ideology of the Animal Rights Movement* (Philadelphia: Temple University Press, 1996), 109.

179 *The streets in Marilyn Carroll's suburban neighborhood* The account of the Halloween visit to Marilyn Carroll's home is based on an interview with Carroll published in the March 1997 *FBR Facts,* a publication of the Foundation for Biomedical Research. This incident and other recent instances of harassment were discussed at the 1997 annual meeting of the National Association for Biomedical Research, which the author attended.

182 *"as long as six billion animals are being consumed as food"* Peter Singer, *Ethics into Action: Henry Spira and the Animal Rights Movement"* (Oxford: Rowman & Littlefield, 1998), 139.

183 *"While the American animal movement"* Ibid., 141.

184 *"I was a bit concerned about inviting her"* Neil L. Wilcox, in conversation with the author, September 1998.

185 true *"advocates of animal rights"* Francione, 110.

185 *"once we abandon animal rights idealism"* Ibid., 79.

185 *"the tone of the 1996 march"* Ibid., 33–34.

185 *"The foundation of animal rights theory"* Ibid., 182.

186 *"The trouble with property status for animals"* Ibid., 183.

188 *"We knew that the cat inner ear"* Walsh's description of the surgery was published in the March/April 1997 issue of *FBR News.* The author contacted Walsh following the 1997 NABR meeting and conducted an e-mail conversation about the PETA campaign and its effect on Walsh, McGee, and their personal and professional lives over the course of a few months. All of the quotes on pages 188–192 are from that conversation.

193 *"As a veterinarian"* John Fioramonti, "IACUCs and Me: A Community Vet's Story, *Newsletter of the Johns Hopkins Center for Alternatives to Animal Testing*, Winter 1996, 8.

194 *"While the regulations seem fairly straightforward"* Tim Allen and D'Anna Jensen, "IACUCs and the Search for Alternatives," *Newsletter of the Johns Hopkins Center for Alternatives to Animal Testing*, Winter 1996, 3.

194 *"The U.S. Department of Agriculture"* Ibid., 5.

195 *"most of the negative reaction to IACUC review results"* Bill D. Roebuck, "IACUC Review: An Investigator's Perspective," *Newsletter of the Johns Hopkins Center for Alternatives to Animal Testing*, Winter 1996, 9.

196 *"Intellectual engagement is the key"* Ibid., p. 10.

196 *"researchers don't see why"* Robert Finn, "Veterinarians in Research Labs Address Conflicting Agendas" *Scientist*, May 26, 1997, 4.

197 *"The ALF . . . thinks you're a pimp"* Ibid.

197 *"The variation in ovary weight"* Michael R. A. Chance and W. M. S. Russell, "The Benefits of Giving Experimental Animals the Best Possible Environment," in Viktor Reinhardt, ed., *Comfortable Quarters for Laboratory Animals* (Washington, D.C.: The Animal Welfare Institute, 1997), 12.

197 *"In both cases, Chance found notable effects"* Ibid.

198 *"Although regulations continue to proliferate"* Linda C. Cork,

Thomas B. Clarkson, Robert O. Jacoby, et al. "The Costs of Animal Research: Origins and Options," *Science*, May 2, 1997, 758.

199 *"When PHS policy and AWRs require appointment"* J. Wesley Robb, "Personal Reflections: The Role and Value of the Unaffiliated Member and the Nonscientist Member of the Institutional Animal Care and Use Committee," *ILAR* [Institute for Laboratory Animal Resources] *News*, vol. 35, no. 3–4: 50.

201 *"despite the impact of the animal rights movement"* Shelley Kaplan and Harold A. Herzog, Jr., "Ethical Ideology and Animal Rights Activism" (unpublished paper, 1994).

203 *"the language of moral crusades"* Jasper and Nelkin, 8.

203 *"The animal liberation/rights movement (ALRM)"* Richard P. Vance, "An Introduction to the Philosophical Presuppositions of the Animal Rights Movement," *Journal of the American Medical Association*, vol. 268, no. 13 (1992): 1715.

203–204 *"Both are exceptionally good philosophers"* Ibid., 1718.

205 *"I find it unfortunate that people in this country"* Erik Parens, conversation with the author, April 1999, following lecture delivered at the University of Maryland.

9: ENGINEERING LIFE

208 *"We know so little about the human body"* W. French Anderson, in Gregory Stock and Robert Campbell, *Engineering the Human Germline Symposium: Summary Report,* draft distributed at the Second Annual Congress on Mammalian Cloning, Washington, D.C., June 26–27, 1998.

210 *"Last summer, biotech food emerged"* Michael Pollan, "Playing God in the Garden," *The New York Times Magazine,* October 25, 1998.

212 *"I still have trouble believing that Dolly is Dolly"* The author attended the First Annual Congress on Mammalian Cloning, in June 1997, and the Second, in June 1998. The comments attributed to various speakers at these meetings are from her notes.

213 *Some, like Princeton molecular biologist Lee M. Silver and philosopher George Ennenga* See Lee M. Silver, *Remaking Eden: Cloning and Beyond in a Brave New Era* (New York: Avon) 1997; George R. Ennenga, "Artificial Evolution" in *Artificial Life* (Boston: MIT Press) vol. 3, no. 1, 1997.

213 *the technique used to produce Dolly* I. Wilmut, A. E. Schnieke, J. McWhir, et al. "Viable Offspring Derived from Fetal and Adult Mammalian Cells," *Nature* 385 (1997): 810–13.

214 *"They killed a lot of embryos"* Cited in John Carey, Naomi Freundlich, Julia Flynn, and Neil Gross, "The Biotech Century," *Business Week,* March 10, 1997.

215 *"The instant I saw the picture"* James D. Watson, *The Double Helix* (New York: Penguin, 1969), 107.

216 *Beginning in 1961, scientists began to explore* For an account of the development of molecular biology and the scientists who built the discipline, see Horace Freeland Judson, *The Eighth Day of Creation* (New York: Simon and Schuster, 1979).

218 *"The Berg experiments scare the pants off a lot of people"* Nicolas Wade, "Microbiology: Hazardous Profession Faces New Uncertainties," *Science,* November 9, 1973, 566.

218 *"scientific developments over the past two years"* D. Singer and D. Soll, "Guidelines for DNA Hybrid Molecules," *Science,* September 21, 1973, 1114.

218–219 *The committee produced the famous Moratorium letter* P. Berg, D. Baltimore, H. W. Boyer, et al. "Potential Hazards of Recombinant DNA Molecules," *Science,* July 26, 1974, 991.

219 *"Increasingly during 1973 we began to ask"* James D. Watson and John Tooze, *The DNA Story: A Documentary History of Gene Cloning"* (San Francisco: W. H. Freeman & Company, 1981), prologue.

220 *"the cloning of recombinant DNA's derived from highly pathogenic organisms"* P. Berg, D. Baltimore, S. Brenner, et al. "Asilomar Conference on Recombinant DNA Molecules," *Science,* June 6, 1975, 991.

221 *"Since the risks and dangers of these technologies"* Letter reproduced in Watson and Tooze, 49.

222 *"there is some importance in genetic experimentation"* Ibid., 97.

222 *"It was quite clear that as soon as science was almost completely funded by government"* Arthur Lubow, "Playing God with DNA," *New Times,* January 7, 1977. Reprinted in Watson and Tooze, 126.

222 *"the evaluation of benefits and risks"* Watson and Tooze, 24.

223 *"Up to now living organisms"* George Wald, "The Case Against Genetic Engineering," *The Sciences* 16 (1976): 6. Reprinted in Watson and Tooze, 25.

223 *the U.S. Supreme Court had ruled that recombinant microorganisms could be patented* The "oncomouse," a mouse genetically altered to be susceptible to cancer, was patented in the United States by Harvard researchers in 1988. The European Patent Office initially rejected the application. After Harvard appealed

the decision, the EPO agreed to issue a patent, but this decision was then appealed by environmental and animal protection groups. However, the patent holds in the United States based on the 1980 Supreme Court ruling, in *Diamond v. Chakrabarty,* that a genetically altered bacteria was patentable. For a discussion of the ethics of patenting life-forms, see Orlans, Beauchamp, Dresser, Morton, and Gluck, eds., *The Human Use of Animals: Case Studies in Ethical Choice* (Oxford: Oxford University Press, 1998).

225 *Reports were issued by advisory groups* Advisory Group on the Ethics of Xenotransplantation, "Animal Tissue Into Humans: A Summary of the Main Conclusions and Recommendations of the Report," August 1996. Available through the U.K. Department of Health, P.O. Box 410, Wetherby, LS23 7LN; Institute of Medicine, *Xenotransplantation: Science, Ethics, and Public Policy* (Washington, D.C.: National Academy Press, 1996).

226 *"virologist Robin Weiss at the Institute for Cancer Research"* Nigel Williams, "Pig-Human Transplants Barred for Now" *Science,* 27 (1997) January 24, 1997, 473.

226 *Meanwhile, animal protectionists argued* Alan Berger and Gill Lamont, "Plunging Headlong into Madness," *Animal Issues* 29(2) (1998): 21–26.

226 *Steven L. Stice, chief scientific officer at Advanced Cell Technologies* W. M. Zawada, J. B. Cibelli, P. K. Choi, et al. "Somatic Cell Cloned Transgenic Bovine Neurons for Transplantation in Parkinsonian Rats," *Nature Medicine* 4, no. 5 (1998).

230 *"How far will we want to develop genetic engineering?"* Robert Sinsheimer, "Troubled Dawn for Genetic Engineering," *New Scientist,* October 16, 1975.

231 *the first death of a gene therapy patient* Sheryl Gay Stolberg, "The Biotech Death of Jesse Gelsinger," *The New York Times Magazine,* November 28, 1999.

232 *"Most traits of medical relevance do not follow simple Mendelian monogenic inheritance"* Eric S. Lander and Nicholas J. Schork, "Genetic Dissection of Complex Traits," *Science,* September 30, 1994, 30.

233 *"it signals a move to more and more testing"* Erik Parens, "Ethics and Genetics: From Testing Ourselves to Designing Our Children," lecture presented at the University of Maryland, April 1999.

233 *"I was overwhelmed by the enormity of the kinds of decisions"* Charles L. Bosk, *All God's Mistakes: Genetic Counseling in a Pediatric Hospital* (Chicago: University of Chicago Press, 1992) xviii.

234 *"First, there's the misinformation argument"* Parens, "Ethics and Genetics" lecture, 1999.

235 *"Medical advantages are not just scientific and technological developments"* Robert Esbjornson, ed. *The Manipulation of Life: Nobel Conference XIX* (San Francisco: Harper & Row, 1984), xiii.

235 *"developments in the medical sciences"* Susan E. Lederer, *Subjected to Science: Human Experimentation in America Before the Second World War* (Baltimore: Johns Hopkins University Press, 1995), 2.

236 *"Physician-researchers viewed the Nuremberg Code as confining"* George J. Annas, "The Changing Landscape of Human Experimentation," in *Medicine, Ethics and the Third Reich*, John Michalczyk, ed. (Kansas City: Sheed & Ward, 1994), 110.

237 *"I'm not sure informed consent is always necessary"* Daniel Koshland, in Gregory Stock and Robert Campbell, *Engineering the Human Germline: Summary Report*, distributed at the Second Annual Congress on Mammalian Cloning, Washington, D.C., June 26–27, 1998, 15.

239 *"Nobody wants to have an average child"* Ibid., 17.

239 *"each child is by nature valuable"* J. Keith Miller, *Compelled to Control* (Deerfield Beach, Fla.: Health Communications, 1997), 63–64.

240 *"Few causes or crusades have such universal support"* Daniel Callahan, "Accepting That Medicine Can Never Erase All Suffering," *Baltimore Sun*, April 15, 1998.

241 *"Scratch the surface of both the information and biotech revolutions"* David Shenk, "Biocapitalism: What Price the Genetic Revolution?" *Harper's Magazine*, December 1997.

10: ANIMALS, SCIENCE, AND THE BODY

242 *"Rationality, when conceived as complete"* Arthur O. Lovejoy, *The Great Chain of Being: A Study of the History of an Idea"* (Cambridge, Mass: Harvard University Press, 1936), 331.

244 *"Foucault seeks to show"* Patrick H. Hutton, "Foucault, Freud and the Technologies of the Self," in Luther H. Martin, Hank Gutman, and Patrick H. Hutton, eds., *Technologies of the Self: A Seminar with Michel Foucault* (Amherst: University of Massachusetts Press, 1988), 127.

245 *"Medicine must no longer be confined to a body of techniques"* Michel Foucault, *The Birth of the Clinic: An Archaeology of*

Medical Perception (New York: Vintage Books/Random House, 1963), 34.

245 *"From the heart of the Middle Ages"* Michel Foucault, "The Politics of Health in the Eighteenth Century," in Colin Gordon, ed., *Power/Knowledge: Selected Interviews & Other Writings 1972–1977* (New York: Pantheon Books, 1972), 170.

246 *"Power would be a fragile thing if its only function were to repress"* Michel Foucault, "Body/Power," in *Power/Knowledge*, 59.

246 *"Where power works 'from below,' prevailing forms of selfhood"* Susan Bordo, *Unbearable Weight: Feminism, Western Culture and the Body"* (Berkeley: University of California Press, 1993), 27.

247 *"The soul is the prison of the body"* Michel Foucault, *Discipline and Punish,* translated from the French by Alan Sheridan (New York: Vintage Books, 1995), 30.

247 *"as a form of power, subjection is paradoxical"* Judith Butler, *The Psychic Life of Power: Theories in Subjection"* (Stanford, Calif.: Stanford University Press, 1997), 2.

248 *"Since the seventeenth century, science has 'owned' the study of the body"* Bordo, 66.

249 *"science 'has all of us by the throat' "* Paul R. Gross and Norman Levitt, *Higher Superstition: The Academic Left and Its Quarrels With Science* (Baltimore: Johns Hopkins University Press, 1994), 27.

249 *"The clinician's gaze"* Foucault, *Birth of the Clinic*, 120.

249 *An article by Malcolm Gladwell on the Ritalin controversy* Malcolm Gladwell, "Running from Ritalin," *The New Yorker*, February 15, 1999.

250 *"The central ambition of the cultural constructivist program"* Gross and Levitt, 69.

250 *"Over all these endeavors"* Foucault, *Birth of the Clinic*, 115.

251 *"A serious investigation of the interplay of cultural and social factors"* Gross and Levitt, 235.

252 *"One now sees the visible only because one knows the language"* Foucault, *Birth of the Clinic*, 115.

253 *"In the last ten years we have come to realize"* Nicolas Wade, "Animal's Genetic Program Decoded, in a Science First," *The New York Times*, December 11, 1998.

256 *A group of women stands in a circle of trees* This ritual is congruent with contemporary Wiccan practices. See Starhawk, *The Spiral Dance: A Rebirth of the Ancient Religion of the Great Goddess* (San Francisco: Harper San Francisco, 1979) and

Zsuzsanna Budapest, *The Holy Book of Women's Mysteries*, (Oakland, Calif.: Wingbow Press, 1980).

258 *"And now I show you a mystery and a new thing"* Anna Kingsford, fragment from *A Prophecy of the Soul, Mystically Called the Day of Woman*, cited in Maitland, 345.

258 *Karen J. Warren, a prominent ecofeminist theorist* Karen J. Warren, "A Feminist Philosophical Perspective on Ecofeminist Spiritualities," in *Ecofeminism and the Sacred*, Carol J. Adams, ed. (New York: Continuum, 1995), 2.

259 *"the same dominant mind-set"* Carol J. Adams, in *Ecofeminism and the Sacred* (New York: Continuum, 1995), 2.

259 *"modern science, which in one way asserts"* Susan Griffin, "Curves Along the Road," in *Reweaving the Web of the World: The Emergence of Ecofeminism*, Irene Diamond and Gloria Feman Orenstein, eds. (San Francisco: Sierra Club Books, 1990), 87.

259 *"It could be argued that theorizing about animals"* Carol J. Adams and Josephine Donovan, eds., *Animals & Women: Feminist Theoretical Explanations* (Durham, N.C.: Duke University Press, 1995), 3.

259–260 *"a broader feminism, a radical cultural feminism"* Adams and Donovan, 3.

260 *"We believe that all forms of oppression"* Ibid., 3.

260 *Their rationale was that women working to eradicate suffering* French, 242.

260 *Clearly, this point of view continues to resonate with many women* See, for example, Riane Eisler, *The Chalice and the Blade: Our History, Our Future* (San Francisco: Harper San Francisco, 1987).

261 *"Numerous studies have established"* Kenneth J. Shapiro, "Attitudes Among Students," in Bekoff, 82–83.

261 *"the duel between the Christian biblical view of the world"* Rosemary Radford Reuther, *Gaia and God: An Ecofeminist Theology of Earth Healing* (San Francisco: Harper San Francisco, 1992), 35.

262 *"all subjective matters of inward experience"* Reuther, 38.

262 *the "triumphalistic assumption of science"* Ibid., 36.

262 *a new dialogue between science and religion* Ibid., 40.

263 *"there are two places to find wisdom"* Matthew Fox, *Original Blessing: A Primer in Creation Spirituality* (Sante Fe, N.M.: Bear and Company, 1983), 9.

263 *"Ecumenism is not about talking together"* Fox, 215.

264 "Turning to the question of 'scientific' vs. 'magical' thinking" Margot Adler, *Drawing Down the Moon: Witches, Druids, Goddess-Worshippers, and Other Pagans in America Today.* (New York: Penguin, 1979), 396–397.

265 *"It is in literature and poetry that we first begin to encounter a reaction"* Gross and Levitt, 20.

266 *It is worth remembering that the Inquisition burned early scientists* See Carlo Ginzburg, *The Cheese and the Worms: The Cosmos of a Sixteenth-Century Miller* (Baltimore: Johns Hopkins University Press, 1980), and Carolyn Merchant, *The Death of Nature: Women, Ecology and the Scientific Revolution* (New York: Harper & Row, 1980).

267 *groups have sometimes formed alliances against their perceived common enemy* For a look at one such alliance, see Coral Lansbury, *The Old Brown Dog: Women, Workers and Antivivisection in Edwardian England* (Madison: University of Wisconsin Press, 1985). The November 1999 riots in the city of Seattle, which disrupted the deliberations of the World Trade Organization, are another example.

267 *"from the Renaissance to the mid-nineteenth century"* W.M.S. Russell, "Biology and Literature in Britain, 1500–1900," part 3, "The Parting of the Ways," *in Social Biology and Human Affairs* 45, no. 1 (1980): 55.

269 *"the love of nature strikes me as essentially composed of democratic passions"* Luc Ferry, *The New Ecological Order*, translated by Carol Volk (Chicago: University of Chicago Press, 1995), 143.

270 *"The following are denounced, in order of their appearance"* Ibid., 69.

271 *"We the undersigned, senior members of the world's scientific community"* Union of Concerned Scientists, *"World Scientists' Warning to Humanity"* (Washington, D.C.: Union of Concerned Scientists, 1993).

271 *"The truth is that we never conquered the world"* Edward O. Wilson, *Biophilia: The Human Bond with Other Species* (Cambridge Mass.: Harvard University Press, 1984), 140.

272 *"this endeavor over the past three centuries"* Thomas Berry, *The Dream of the Earth* (San Francisco: Sierra Club Books, 1988), 29.

272 *"What we need, what we are ultimately groping toward"* Ibid., 48–49.

273 *"Natural philosophy has brought into clear relief"* Wilson, 140.

273 *"the industrial context in which we presently function"* Berry, 30.

273 *"We can do nothing adequate"* Ibid., 69.

273 *a "laissez-faire liberal attitude is nonetheless insane"* Ferry, 127.

274 *"if dominating and destructive relations to the earth"* Reuther, 4–5.

275 *"We know from experience"* Michel Foucault, "What Is Enlightenment?" in *The Foucault Reader*, Paul Rabinow, ed. (New York: Pantheon, 1984), 46–47.

11: PARTIAL TRANSFORMATIONS

276 *"Public policy, of course, is not made in a vacuum"* Robert Garner, *Animals, Politics, and Morality* (Manchester, England: Manchester University Press, 1993), 181.

276 *"The Germans have a word to describe the Dutch temperament"* Otto Postma's comments about the Netherlands and its people were made during a presentation at the Second Annual Congress on Mammalian Cloning, held in Washington, D.C., in June 1998. The author conducted a follow-up conversation with Gerald van Beynum, a colleague of Postma's, later that year.

277 *"The essence of any effective system is dialogue"* Judith Hampson, "Legislation: A Practical Solution to the Vivisection Dilemma," in Rupke, 314–339.

278 *when Wade Roush, a reporter for* Science . . . *attempted to locate a Dutch scientist* The author spoke to Roush about his difficulty finding a scientist opposed to the three R's while she was at the Johns Hopkins Center for Alternatives to Animal Testing.

278 *"People are much more strict"* Wade Roush, "Hunting for Animal Alternatives," *Science*, October 11, 1996, 170.

278 *"We don't always agree, but we don't fight anymore"* Paul de Greeve, during an interview with the author at the Second World Congress on Animal Use and Alternatives in the Life Sciences, Utrecht, October, 1996.

279 *"The consensus-building approach is typically Danish"* Mark Matfield, during an interview with the author in London, July 1998.

280 *"The role of governments with regard to animal experimentation"* Erica Terpstra, opening remarks at the Second World Congress on Animal Use and Alternatives, Utrecht, October 1996.

280 *"With regard to laboratory animals"* Netherlands Center for Alternatives Newsletter, July 1997.

281 *"The three-week course was developed in 1985"* L. F. M. van Zutphen, interview with the author, October 1996, and follow-up conversation in September 1998.

283 *"Animal experimentation is considered something of a poisoned chalice"* Mark Matfield, conversation with the author, London, July 1998. All subsequent quotes in this chapter are from the same interview.

284 *"The Animals (Scientific Procedures) Act 1986 provides specific authority* C. B. Hart, in *Laboratory Animals: An Introduction for the Experimenter,* A. A. Tuffery, ed. (Chichester, England: John Wiley & Sons, 1995), 42.

286 *"There is certainly an increased sympathy and awareness of animal rights issues"* David Dewhurst, lecture at the Johns Hopkins School of Public Health, followed by interview by author, March 1999.

288 *"The state's role, largely confirmed by the evidence"* Garner, plenary lecture delivered at the Second World Congress on Animal Use and Alternatives, Utrecht, October 1996.

288 *"legislative controls on the use of animals"* Garner, Second World Congress lecture.

288–289 *"although its origins lie in a breakaway faction of the HSA"* Garner, *Animals, Politics and Morality,* 218.

289 *"The use of high explosive"* Ibid., 219.

290 *"It has been claimed, for instance"* Ibid., 221.

290 *"The kidnappers put a hood over his head"* Paul Lashmar, "Journalist Seized and Branded by Animal Activists," *The Independent,* November 8, 1999.

291 *prominently featured on the Americans for Medical Progress Web site, AMP News.* (www.ampef.org).

291 *"This is a fight between money and decency"* Alison Schneider, "As Threats of Violence Escalate, Primate Researchers Stand Firm," *The Chronicle of Higher Education,* November 12, 1999, A17.

292 *"There is no doubt that Europeans"* Gillian Griffin, "A Tale of Two Continents," *Newsletter of the Johns Hopkins Center for Alternatives to Animal Testing,* vol. 12 (3), 6.

292 *"because of variations in detail"* Hart, 63.

293 *Mark Matfield* Interview with the author, July 1998.

295 *"it was difficult to persuade the agency"* Neil Wilcox, interview with the author, Washington, D.C., September 1998.

296 *"When the profession can no longer evade anomalies"* Thomas Kuhn, *The Structure of Scientific Revolutions* (Chicago: University of Chicago Press, 1970), 82–83.

296 *"The proponents of competing paradigms"* Ibid., 148–50.

296 *critics of whole-animal methods* Joanne Zurlo, Deborah Ruda-cille, and Alan M. Goldberg, *Animals and Alternatives in Testing: History, Science and Ethics* (New York: Mary Ann Liebert, 1993), 21.

298 *"In other words, the public cannot tell"* David Roe, *Toxic Igno-rance* (Washington, D.C.: Environmental Defense Fund, 1997).

300 *"Significant doubt exists about the validity of animal testing results"* Ken Calvert, House Subcommittee Hearing, June 17, 1999. The author attended the hearing, and all remarks quoted are from her notes or the written record of the hearing.

301 *"for personal reasons"* Hill accompanied William Saunders to the June 1999 Capitol Hill hearing and sat at the witness table with him, though all questions directed to the EPA about the HPV program were answered by Saunders. The author spoke briefly to Hill after the hearing.

303 *"NIH does officially support the three R's"* Louis Sibal, interview with the author, March 1999.

304 *"there is a $40 billion research industry"* Andrew Rowan, inter-view with the author, March 1999.

305 *"I can remember giving a talk thirty years ago"* Thomas Wolfle, presentation at a workshop sponsored by the Scientists Center for Animal Welfare, reprinted in *The Human/Research Animal Relationship* (a compilation of presentations from three work-shops held on the subject from 1992 to 1994), Lee Krulisch, ed. (Greenbelt, Md.: Scientists Center for Animal Welfare, 1996), 89.

306 *"Some principal investigators reported to Dr. Carmack"* Ibid., 36.

307 *"All of the coping mechanisms I have described"* Ibid., 44.

307 *"I felt there a tremendous fear from participants"* Joy Becker, question-and-answer session following Kristina Stephens's talk "Animal Staff Perspectives," in *The Human/Research Animal Relationship*, 64.

307–308 *"At Bryn Mawr, my early science training"* Candace Pert, *Molecules of Emotion* (New York: Scribner's, 1997), 34.

308 *"I knew that I was going to have to desensitize myself"* Ibid., 52.

308 *"Regardless of how ethical"* Spinelli, "Human/Research Animal Relationships: Animal Staff Perspectives," in *The Human/Research Animal Relationship*, 47.

309 *"I was told that my choice of title"* Arnold Arluke, "The Well-Being of Animal Researchers," ibid., 7.

310 *early in 1999 a request appeared on COMPMED* Andrew Rowan, in conversation with the author, March 1999.

311 *"Many friends and colleagues have urged me not to confess my feelings"* Edward Walsh, in conversation with the author, 1999.

312 *"Most people want to keep a low profile"* Arluke, 14.

EPILOGUE

314 *Mr. Goldstein, who has been charged with second-degree murder* Michael Cooper, "Man Accused in Train Death Says He Tried to Attack Before," *The New York Times*, January 6, 1999.

314 *"Schizophrenia destroys what is distinctly human"* William T. Carpenter, in conversation with the author and members of the University of Maryland Office of Communication and Publications, February 1999.

315 *"First, you're shocked"* Brett Levinson, University of Maryland medical student, interviewed by author, January 1999.

317 *"has no specific policy on research involving persons with mental disabilities"* Rebecca Dresser, "Mentally Disabled Research Subjects: The Enduring Policy Issues," *Journal of the American Medical Association* 276 (1996): 67–72. See also Robert J. Levine, "Proposed Regulations for Research Involving Those Institutionalized as Mentally Infirm: A Consideration of Their Relevance in 1995," in *Accountability in Research*, vol. 4 (Amsterdam: Gordon & Breach Science Publishers, 1996), 177–86.

317 *"Though current protection may be inadequate"* Ibid., 67.

319 *"Generally, we refuse to admit within ourselves"* Joseph Campbell, *The Hero with a Thousand Faces* (Princeton: Princeton University Press, 1949), 121.

319 *"The battlefield is symbolic of the field of life"* Ibid., 238.

320 *"Nothing retains its own form"* Ovid, cited in Campbell, p. 243.

320 *"Not the animal world, not the plant world"* Ibid., 391.

REFERENCES

The following works were used in the making of this book. Most are cited in the text; others provided background information or context for the issues discussed here. This is not a complete record of all the works and sources I have consulted, but instead indicates the range of viewpoints and authors that have informed my views on this subject.

Adams, Carol J., ed. *Ecofeminism and the Sacred*. New York: Continuum, 1995.

Adams, Carol J. *Neither Man nor Beast: Feminism and the Defense of Animals*. New York: Continuum, 1995.

Adams, Carol J., and Josephine Donovan, eds. *Animals & Women: Feminist Theoretical Explorations*. Durham and London: Duke University Press, 1995.

Adler, Margot. *Drawing Down the Moon: Witches, Druids, Goddess-Worshippers, and Other Pagans in America Today*. New York: Penguin, 1979.

Advisory Group on the Ethics of Xenotransplantation. "Animal Tissue Into Humans." London: Department of Health, 1996.

Allen, Tim, and D'Anna Jensen. "IACUCs and AWIC: The Search for Alternatives." *Johns Hopkins Center for Alternatives to Animal Testing Newsletter*, vol. 13 (Winter 1996).

Animal Welfare Institute. *Animals and their Legal Rights: A Survey of American Laws from 1641 to 1990*. Washington, D.C.: Animal Welfare Institute, 1968.

———. "Information Report." Washington, D.C.: Animal Welfare Institute, 1952–1962.

References

Annas, George. "The Changing Landscape of Human Experimentation." In *Medicine, Ethics and the Third Reich: Historical and Contemporary Issues*, edited by John J. Michalczyk. Kansas City, Mo.: Sheed & Ward, 1994.

Arluke, Arnold. "The Well-being of Animal Researchers." In *The Human/Research Animal Relationship*, edited by Lee Krulisch. Greenbelt, Md.: Scientists Center for Animal Welfare, 1996.

Arluke, Arnold, and Clinton R. Sanders. *Regarding Animals*. Philadelphia: Temple University Press, 1996.

Bentham, Jeremy. *Introduction to the Principle of Morals and Legislation*. 1780. Reprint, London: Macmillan, 1963.

Berger, Alan, and Gil Lamont. "Plunging Headlong into Madness." *Animal Issues*, vol. 29 (Summer 1998).

Bernard, Claude. *An Introduction to the Study of Experimental Medicine*. Translated by Henry Copley Green. New York: Macmillan, 1927.

Berg, P., D. Baltimore, H. W. Boyer, et al. "Potential Biohazards of Recombinant DNA Molecules." *Science,* July 26, 1974, 303.

Berg, P., D. Baltimore, S. Brenner, et al. "Asilomar Conference on Recombinant DNA Molecules." *Science,* June 6, 1975, 991.

Berry, Thomas. *The Dream of the Earth*. San Francisco: Sierra Club Books, 1988.

Blum, Deborah. *The Monkey Wars*. New York and Oxford: Oxford University Press, 1994.

Bohlen, Celeste. "Animal Rights Case: Terror or Entrapment?" *The New York Times*, March 3, 1989.

Bordo, Susan. *Unbearable Weight: Feminism, Western Culture and the Body*. Berkeley: University of California Press, 1993.

Bosk, Charles L. *All God's Mistakes: Genetic Counseling in a Pediatric Hospital*. Chicago and London: University of Chicago Press, 1992.

Bronowski, Jacob. *Science and Human Values*. New York: Harper and Row, 1956.

Brooks, Linda. "U.S. Surgical Handling Crisis." *Norwalk (Conn.) Hour*, February 14, 1989.

Butler, Judith. *The Psychic Life of Power: Theories in Subjection*. Stanford, Calif.: Stanford University Press, 1997.

Callahan, Daniel. "Accepting That Medicine Can Never Erase All Suffering." *Baltimore Sun*, April 15, 1998.

Campbell, Joseph. *The Hero with a Thousand Faces*. Princeton: Princeton University Press, 1949.

Capsis, John. "Feds Investigate Trutt Sting." *Westport (Conn.) News*, February 1, 1989.

References

———. "Police, U.S. Surgical Implicated in Bomb Plot: Informer Says He Was Paid by U.S. Surgical to Transport Bomb." *Westport (Conn.) News*, January 11, 1989.

———. "Police Confirm Operative Delivered Trutt and Bomb." *Westport (Conn.) News*, January 13, 1989.

———. "Trutt: Mead Gave Me $1,200 to Buy the Bomb." *Westport (Conn.) News*, January 16, 1989.

———. " 'Trust Mead' Trutt Told by Second Operative." *Westport (Conn.) News*, January 25, 1989.

Carey, John, Naomi Freundlich, Julia Flynn, and Neil Gross. "The Biotech Century." *Business Week*, February 27, 1997.

Chance, Michael R. A. and W. M. S. Russell. "The Benefits of Giving Experimental Animals the Best Possible Environment." In *Comfortable Quarters for Laboratory Animals*, edited by Viktor Reinhardt. Washington, D.C.: The Animal Welfare Institute, 1997.

Cobbe, Frances Powers. *Life of Frances Powers Cobbe as Told by Herself*. London: Swan, Sonnenschein, 1904.

Cobbe, Frances Powers. *The Modern Rack*. London: Swan, Sonnenschein, 1889.

Cooney, Beth. "Trutt Sentenced to 32 Months in Prison." *Westport (Conn.) Advocate*, July 17, 1990.

Cork, Linda C., Thomas B. Clarkson, Robert O. Jacoby, et al. "The Costs of Animal Research: Origins and Options." *Science,* May 2, 1997, 758–59.

De Kruif, Paul. *Microbe Hunters*. New York: Harcourt, Brace, 1926.

Dodds, W. J. and F. B. Orlans. *Scientific Perspectives on Animal Welfare*. New York: Academic Press, 1982.

Dresser, Rebecca. "Mentally Disabled Research Subjects: The Enduring Policy Issues." Journal of the American Medical Association, 276, no. 1 (1996): 67–72.

Driscoll, J. "Attitudes Toward Animal Use." *Anthrozoos*, vol. 5 (1992).

Duster, Troy. *Backdoor to Eugenics*. New York and London: Routledge, 1990.

Eisler, Riane. *The Chalice and the Blade: Our History, Our Future*. San Francisco: Harper San Francisco, 1987.

Elston, Mary Ann. "Women and Antivivisection in Victorian England." In *Vivisection in Historical Perspective*, edited by Nicolaas Rupke. London and New York: Routledge, 1987.

Ennenga, George R. "Artificial Evolution." *Artificial Life* 3, no. 1 (1997): 51–61.

Esbjornson, Robert, ed. *The Manipulation of Life: Nobel Conference XIX*. San Francisco: Harper & Row Publishers, 1984.

Fee, Elizabeth. *Disease and Discovery: A History of the Johns Hopkins School of Public Health 1916–1939*. Baltimore and London: Johns Hopkins University Press, 1987.

Ferry, Luc. *The New Ecological Order*. Translated by Carol Volk. Chicago and London: University of Chicago Press, 1995.

Finn, Robert. "Veterinarians in Research Labs Address Conflicting Agendas. *Scientist*, May 26, 1997.

Finsen, Susan, and Lawrence Finsen. "Animal Rights Movement." In *The Encyclopedia of Animal Rights and Animal Welfare*, edited by Marc Bekoff with Carron A. Meaney. Westport, Conn.: Greenwood Press, 1998.

Fioramonti, John. "IACUCs and Me: A Community Vet's Story." *Johns Hopkins Center for Alternatives to Animal Testing Newsletter*. vol. 13 (Winter 1996).

Flexner, Abraham. *Medical Education in the United States and Canada: A Report to the Carnegie Foundation for the Advancement of Teaching*. New York: Carnegie Foundation, 1910.

Flexner, Simon, and James Thomas Flexner. *William Henry Welch and the Heroic Age of American Medicine*. New York: The Viking Press, 1941.

Foucault, Michel. *The Birth of the Clinic: An Archaeology of Medical Perception*. Translated by A. M. Sheridan Smith. New York: Vintage Books/Random House, 1963.

———. "The Political Technology of Individuals." In *Technologies of the Self: A Seminar with Michel Foucault*, edited by Luther H. Martin, Hank Gutman, and Patrick H. Hutton. Amherst, Mass.: University of Massachusetts Press, 1988.

———. *Power/Knowledge: Selected Interviews and Other Writings 1972–1977*, edited and translated by Colin Gordon. New York: Pantheon Books, 1980.

Fox, Matthew. *Original Blessing: A Primer in Creation Spirituality*. Sante Fe, N.M.: Bear and Company, 1983.

Francione, Gary. *Rain Without Thunder: The Ideology of the Animal Rights Movement*. Philadelphia: Temple University Press, 1996.

French, Richard D. *Antivivisection and Medical Science in Victorian Society*. Princeton: Princeton University Press, 1975.

Galvin, S., and H. Herzog. "Ethical Ideology, Animal Rights Activism, and Attitudes Toward the Treatment of Animals." *Ethics and Behavior*, vol. 2 (1992).

Garner, Robert. *Animals, Politics and Morality*. Manchester, England, and New York: Manchester University Press, 1993.

Gesell, Robert. Letters to A. J. Carlson from the files of the Animal Welfare Institute, Washington, D.C., 1946–1952.

Ginzburg, Carlo. *The Cheese and the Worms: The Cosmos of a Sixteenth-Century Miller.* Baltimore and London: Johns Hopkins University Press, 1980.

Gladwell, Malcolm. "Running from Ritalin." *The New Yorker*, February 15, 1999.

Glass, James M. *Life Unworthy of Life: Racial Phobia and Mass Murder in Hitler's Germany.* New York: Basic Books, 1997.

Gould, Stephen Jay. *Ever Since Darwin: Reflections in Natural History.* New York and London: W. W. Norton, 1977.

Greenspan, Patricia S. *Emotions & Reasons: An Inquiry into Emotional Justification.* New York and London: Routledge, 1988.

Griffin, Gillian. "A Tale of Two Continents." *Johns Hopkins Center for Alternatives to Animal Testing Newsletter*, vol. 12 (Summer 1995).

Griffin, Susan. "Curves Along the Road." In *Reweaving the World: The Emergence of Ecofeminism*, edited by Irene Diamond and Gloria Feman Orenstein. San Francisco: Sierra Club Books, 1990.

Griswold, Charles W., Jr. "Ingrid Newkirk: Q and A." *Washington City Paper*, December 20, 1985.

Gross, Paul R., and Norman Levitt. *Higher Superstition: The Academic Left and Its Quarrels with Science.* Baltimore and London: Johns Hopkins University Press, 1994.

Groves, Julian McAllister. *Hearts and Minds: The Controversy over Laboratory Animals.* Philadelphia: Temple University Press, 1997.

Hampson, Judith. "Legislation: A Practical Solution to the Vivisection Dilemma." In *Vivisection in Historical Perspective*, edited by Nicolaas Rupke. London and New York: Routledge, 1990.

Hanauske-Abel, Hartmut. "Not a Slippery Slope or a Sudden Subversion: German Medicine and National Socialism in 1993." *British Medical Journal*, December 7, 1996, 1453–63.

Hardin, Victoria A. *Inventing the NIH: Federal Biomedical Research Policy 1887–1937.* Baltimore and London: Johns Hopkins University Press, 1986.

Harre, Rom. *Great Scientific Experiments: 20 Experiments That Changed Our View of the World.* Oxford: Phaidon, 1981.

Harrington, Anne. *Reenchanted Science: Holism in German Culture from Wilhelm II to Hitler.* Princeton, N.J.: Princeton University Press, 1996.

Hart, C. B. *Laboratory Animals: An Introduction for the Experimenter.* Chichester, England: John Wiley & Sons, 1995.

References

Herzog, Harold A., Jr. "Sociology of the Animal Rights Movement." In *The Encyclopedia of Animal Rights and Animal Welfare*, edited by Marc Bekoff with Carron A. Meaney. Westport, Conn.: Greenwood Press, 1998.

Hoedeman, Paul. *Hitler or Hippocrates: Medical Experiments and Euthanasia in the Third Reich*. Translated from the Dutch by Ralph de Rijke. Sussex, England: The Book Guild, 1991.

Hume, E. D. *The Mind-Changers*. London: Michael Joseph, 1939.

Hutton, Patrick. "Foucault, Freud and the Technologies of the Self." In *Technologies of the Self: A Seminar with Michel Foucault*, edited by Luther H. Martin, Hank Gutman, and Patrick H. Hutton. Amherst Mass.: University of Massachusetts Press, 1988.

Institute of Medicine. "Xenotransplantation: Science, Ethics and Public Policy." Washington, D.C.: National Academy Press, 1996.

Jasper, James, and Dorothy Nelkin. *The Animal Rights Crusade: The Growth of a Moral Protest*. New York: The Free Press, 1992.

Jones, James H. *Bad Blood: The Tuskegee Syphilis Experiment*. New York: The Free Press, 1981.

Judson, Horace Freeland. *The Eighth Day of Creation: The Makers of the Revolution in Biology*. New York: Simon & Schuster, 1979.

Kaufman, Ron. "Scientists Doubtful About New Law Aiming to Protect Animal Research Facilities." *The Scientist*, October 26, 1992.

Kessler, Robert E., and Michael Slackman. "Informant Admits Role in Bombing." *New York Newsday*, January 12, 1989.

Kevles, Daniel J. *In the Name of Eugenics: Genetics and the Uses of Human Heredity*. New York: Alfred A. Knopf, 1985.

Klein, Aaron E. *Trial by Fury: The Polio Vaccine Controversy*. New York: Charles Scribner's Sons, 1972.

Knight, D. *The Age of Science: The Scientific Worldview in the Nineteenth Century*. London: Basil Blackwell, 1986.

Kolata Gina. *Clone: The Road to Dolly and the Path Ahead*. New York: William Morrow, 1998.

———. "Scientists Urge Senators Not to Rush Ban on Human Cloning." *The New York Times*, March 13, 1997.

Kuhn, Thomas S. *The Structure of Scientific Revolutions*. Chicago: University of Chicago Press, 1970.

Lamb, Ruth deForest. *American Chamber of Horrors: The Truth About Food and Drugs*. New York: Farrar and Rinehart, 1926.

Lander, Eric S., and Nicholas J. Schork. "Genetic Dissection of Complex Traits." *Science*, September 30, 1994.

Lansbury, Coral. *The Old Brown Dog: Women, Workers and Vivisection in Edwardian England.* Madison: University of Wisconsin Press, 1985.

Lansdell, Herbert C. "The Three Rs: A Restrictive and Refutable Rigamarole." In *Ethics and Behavior,* vol. 3, no. 2 (1993): 177–185.

Lederer, Susan E. "Controversy in America, 1880–1914." In *Vivisection in Historical Perspective,* edited by Nicolaas Rupke. London and New York: Routledge, 1987.

———. *Subject to Science: Human Experimentation in America Before the Second World War.* Baltimore and London: Johns Hopkins University Press, 1995.

Lifton, Robert J. *The Nazi Doctors: Medical Killing and the Psychology of Genocide.* New York: Basic Books, 1986.

Lindsey, J. R. "Historical Foundations." In *The Laboratory Rat,* vol. 1, edited by H. J. Baker, J. R. Lindsey, and S. H. Weisbroth. New York: Academic Press, 1979.

Linzey, Andrew. "Brigid Brophy." In *The Encyclopedia of Animal Rights and Animal Welfare,* edited by Marc Bekoff with Carron A. Meaney. Westport, Conn.: Greenwood Press, 1998.

Lovejoy, Arthur O. *The Great Chain of Being: A Study of the History of an Idea.* Cambridge and London: Harvard University Press, 1936.

Lubow, Arthur. "Playing God with DNA." *New Times,* January 7, 1977, 48.

Maehle, Andreas-Holger, and Ulrich Troehler. "Animal Experimentation from Antiquity to the End of the Eighteenth Century." In *Vivisection in Historical Perspective,* edited by Nicolaas Rupke. London and New York: Routledge, 1987.

Maitland, Edward. *Anna Kingsford: Her Diaries, Letters, and Work.* London: Watkins, 1896.

Mangan, Katherine S. "Universities Beef Up Security at Laboratories to Protect Researchers Threatened by Animal Rights Activists." *Chronicle of Higher Education,* September 19, 1990, A16.

McCrea, Roswell C. *The Humane Movement: A Descriptive Survey.* New York: Columbia University Press, 1910.

McIntire, Mike. "Trailing of Trutt Apparently One Facet of Medical Researcher's Broader Plan." *Norwalk (Conn.) Hour,* February 2, 1989.

———. "Trutt: Agent Aided Me Every Step of the Way." *Norwalk (Conn.) Hour,* January 27, 1989.

McIntire, Mike, and Ed Silverstein. "Mead's Lawyer Linked to Detective Firm." *Norwalk (Conn.) Hour,* January 12, 1989.

References

Merchant, Carolyn. *The Death of Nature: Women, Ecology and the Scientific Revolution*. New York: Harper and Row, 1980.

Miller, J. Keith. *Compelled to Control*. Deerfield Beach Fla.: Health Communications, 1997.

Morse, Herbert C. "The Laboratory Mouse—An Historical Perspective." In *The Mouse in Biomedical Research*, vol. 1, edited by Henry L. Foster, David Small, and James G. Fox. New York: Academic Press, 1981.

Netherlands Centre for Alternatives to Animal Use. "Dutch Experiments on Animals Act (1977) Revised." *Netherlands Center Alternatives to Animal Use Newsletter*, no. 4 (July 1997).

Newkirk, Ingrid. 1992. *Free the Animals: The Inside Story of the Animal Liberation Front*. Chicago: Noble Press, 1992.

Niven, Charles D. *History of the Humane Movement*. London: Johnson, 1967.

Nyiszli, Miklos. *Auschwitz: A Doctor's Eyewitness Account*. Translated by Tibere Kremer and Richard Seaver. New York: Arcade Publishing, 1960.

Olmstead, E. Harris, and M. D. James. *Claude Bernard and the Experimental Method in Medicine*. New York: Henry Schuman, 1952.

Orlans, Barbara. *In the Name of Science: Issues in Responsible Animal Experimentation*. New York and Oxford: Oxford University Press, 1993.

Orlans, F. B., T. Beauchamp, R. Dresser, D. Morton, and J. P. Gluck. *The Human Use of Animals: Case Studies in Ethical Choice*. New York and Oxford: Oxford University Press, 1998.

Pacheco, Alex, with Anna Francione. "The Silver Spring Monkeys." In *In Defense of Animals*, edited by Peter Singer. New York: Basil Blackwell, 1985.

Palmer, Barclay. "Trutt OK's Plea Agreement." *The Advocate*, April 17, 1990.

Parascandola, John. "The History of Animal Use in the Life Sciences." In *Proceedings of the World Congress on Alternatives and Animal Use in the Life Sciences: Research, Education and Testing*. Vol. II of *Alternative Methods in Toxicology*, edited by Alan M. Goldberg. New York: Mary Ann Liebert, 1994.

Patience, Clive. "Infection of Human Cells by an Endogenous Retrovirus of Pigs." *Nature Medicine* 3 (1997): 282–86.

Paul, John R. *A History of Poliomyelitis*. New Haven and London: Yale University Press, 1971.

Perkins, Judith. *The Suffering Self: Pain and Narrative Representation in the Early Christian Era*. London: Routledge, 1995.

Perry, Luc. *The New Ecological Order*. Chicago and London: University of Chicago Press, 1995.

Pert, Candace. *Molecules of Emotion*. New York: Charles Scribner's Sons, 1997.

Pollan, Michael. "Playing God in the Garden." *The New York Times Magazine*, October 25, 1998.

Proctor, Robert. *Racial Hygiene: Medicine Under the Nazis*. Cambridge: Harvard University Press, 1988.

Rabinow, Paul, ed. *The Foucault Reader*. New York: Pantheon Books, 1984.

Regan, Tom. *The Case for Animal Rights*. Berkeley: University of California Press, 1983.

Regan, Tom, and Peter Singer. *Animal Rights and Human Obligations*. Englewood Cliffs, N.J.: Prentice-Hall, 1976.

Richardson, Ruth. *Death, Dissection and the Destitute*. London: Routledge & Kegan, 1987.

Robb, J. Wesley. "Personal Reflections: The Role and Value of the Unaffiliated Member and the Nonscientist Member of the Institutional Animal Care and Use Committee." *ILAR News*, vol. 35, no. 3–4 (1993).

Roe, David. *Toxic Ignorance*. Washington, D.C.: Environmental Defense Fund, 1997.

Roebuck, Bill D. "IACUC Review: An Investigator's Perspective." *Johns Hopkins Center for Alternatives to Animal Testing Newsletter*, vol. 13, no. 2 (1996).

Rogers, Naomi. *Screen the Baby, Swat the Fly: Polio in the Northeastern U.S., 1916*. Doctoral dissertation, University of Pennsylvania. Ann Arbor, Mich.: Dissertation Information Service, 1986.

Rollin Bernard E. *Animal Rights & Human Morality*. Buffalo, N.Y.: Prometheus Books, 1992.

———. *The Unheeded Cry: Animal Consciousness, Animal Pain and Science*. Oxford: Oxford University Press, 1989.

Rose, Mara, Beth Cooney, and Denise Buffa. "U.S. Surgical Denies Role in Bomb Plot." *The Advocate*, January 14, 1989.

Roush, Wade. "Going Dutch: Mainstreaming Alternatives." *Science*, October 11, 1996, 170.

———. "Hunting for Animal Alternatives." *Science*, October 11, 1996.

Rowan, Andrew. *Of Mice, Models, and Men: A Critical Evaluation of*

Animal Research. Albany: State University of New York Press, 1984.

Rowan, Andrew, Franklin Loew, and Joan Weer. *The Animal Rights Controversy: Protest, Process and Public Policy*. Boston: Tufts University School of Veterinary Medicine, 1995.

Rudacille, Deborah. "NIH Submits 3R's Report to Congress." *Johns Hopkins Center for Alternatives to Animal Testing Newsletter*, vol. 12, no. 1 (1994).

———. "Second World Congress Highlights Policy, Ethics, Refinement." *Johns Hopkins Center for Alternatives to Animal Testing Newsletter*, vol. 14, no. 2 (1997).

Ruether, Rosemary Radford. *Gaia and God: An Ecofeminist Theology of Earth Healing*. San Francisco: HarperSanFrancisco, 1992.

Rupke, Nicolaas A., ed. *Vivisection in Historical Perspective*. London and New York: Routledge, 1987.

Russell, W. M. S. Vol. 3 of *Biology and Literature in Britain, 1500–1900*. "The Parting of the Ways." In *Social Biology and Human Affairs*, vol. 45, no 1 (1980).

Russell, W. M. S., and R. L. Burch. *The Principles of Humane Experimental Technique*. Reprint, Hertfordshire, England: Universities Federation for Animal Welfare, 1992.

Ruth, David N., and Robert W. Cox. *The Care and Management of Laboratory Animals Used in Programs of the Department of Health, Education and Welfare*. Washington, D.C.: Division of Operations Analysis, Office of the Comptroller, Office of the Secretary, 1966.

Scientists Center for Animal Welfare. *The Human/Research Animal Relationship*, edited by Lee Krulisch. Greenbelt, Md.: Scientists Center for Animal Welfare, 1996.

Secher, J. A. *The Role of Animals in Biomedical Research*. New York: Annals of the New York Academy of Sciences, vol. 406, 1983.

Shapiro, Kenneth J. "Attitudes Among Students." In *The Encyclopedia of Animal Rights and Animal Welfare*, edited by Marc Bekoff with Carron A. Meaney. Westport, Conn.: Greenwood Press, 1998.

Shenk, David. "Biocapitalism: What Price the Genetic Revolution?" *Harper's*, December 1997.

Silver, Lee. *Remaking Eden: Cloning and Beyond in a Brave New Era*. New York: Avon Books, 1997.

Singer, D., and D. Soll. "Guidelines for DNA Hybrid Molecules." *Science*, September 21, 1973, 1114.

Singer, Peter. *Animal Liberation*. New York: Avon Books, 1975.

———. *Ethics into Action: Henry Spira and the Animal Rights Movement*. Oxford: Rowman & Littlefield, 1998.

Sinsheimer, Robert. "Troubled Dawn for Genetic Engineering." *New Scientist,* October 16, 1975, 148.

Smith, Jane A. *Patenting the Sun: Polio and the Salk Vaccine.* New York: William Morrow, 1990.

Smith-Rosenberg, Carroll. "The Hysterical Woman: Sex Roles and Role Conflict in Nineteenth Century America." In *The Yellow Wallpaper.* Women Writers: Texts and Contexts Series, Thomas L. Erskine and Connie L. Richards, series eds. New Brunswick: Rutgers University Press, 1992.

Smyth, David H. *Alternatives to Animal Experiments.* London: Scolar Press, 1978.

Sperling, Susan. *Animal Liberators: Research and Morality.* Los Angeles and Berkeley: University of California Press, 1988.

Spinelli, Joseph S. "Human/Research Animal Relationships: Animal Staff Perspectives." In *The Human/Research Animal Relationship,* edited by Lee Krulisch. Greenbelt, Md.: Scientists Center for Animal Welfare, 1996.

Spira, Henry. "Animals Suffer for Science." *Our Town,* July 23, 1976. Reprinted in *Strategies for Activists: From the Campaign Files of Henry Spira,* edited by Henry Spira. New York: Animal Rights International, 1997.

———. "Museum Victory for Animal Rights." *Our Town,* February 26, 1978. Reprinted in *Strategies for Activists: From the Campaign Files of Henry Spira,* edited by Henry Spira. New York: Animal Rights International, 1997.

———. "Rule and Ruin in the NMU." *The Call,* 1969. Reprinted in *Strategies for Activists.*

Stock, Gregory, and Robert Campbell. "Engineering the Human Germline: Summary Report." Draft of proceedings distributed at the Second Annual Congress on Mammalian Cloning. Washington, D.C., June 26–27, 1998.

Taylor, Robert. "A Pig in a Poke: Xenotransplants and Infectious Disease." *Nature Medicine,* vol. 1, no. 8 (August 1995).

Thomas, Keith. *Man and the Natural World: A History of the Modern Sensibility.* New York: Pantheon Books, 1983.

Tierney, Margaret. "Terms of Endearment: The Story of Mary Lou Sapone and Fran Trutt." *Norwalk (Conn.) Hour,* April 21, 1990.

Turner, James. *Reckoning with the Beast: Animals, Pain and Humanity in the Victorian Mind.* Baltimore and London: Johns Hopkins University Press, 1980.

Union of Concerned Scientists. "World Scientists' Warning to Humanity." Washington, D.C.: Union of Concerned Scientists, 1993.

References

U.S. Senate. "Prohibiting Vivisection of Dogs." Senate Committee on the Judiciary, Bill 66 S. 1258, 1919.

Vance, Richard P. "An Introduction to the Philosophical Presuppositions of the Animal Liberation/Rights Movement." *Journal of the American Medical Association,* October 7, 1992.

Vyvyan, John. *In Pity and in Anger: A Study of the Use of Animals in Science.* London: Michael Joseph, 1969.

Wade, Nicholas. "Animal's Genetic Program Decoded, in a Science First." *The New York Times,* December 11, 1998.

———. "Animal Rights: NIH Cat Sex Study Brings Grief to New York Museum." *Science,* October 8, 1976, 162–67.

———. "Microbiology: Hazardous Profession Faces New Uncertainties." *Science,* November 9, 1973, 566.

———. *The Ultimate Experiment: Man-Made Evolution.* New York: Walker and Company, 1977.

Wald, George. "The Case Against Genetic Engineering." *The Sciences* 16 (September 1976): 6.

Warren, Karen J. "A Feminist Philosophical Perspective on Ecofeminist Spiritualities." In *Ecofeminism and the Sacred,* edited by Carol J. Adams. New York: Continuum, 1993.

Watson, James D. *The Double Helix.* New York: Atheneum, 1968.

Watson, James D., with John Tooze. *The DNA Story: A Documentary History of Gene Cloning.* San Francisco: W. H. Freeman & Company, 1981.

Weiss, Rick. "Panel Backs Some Human Clone Work." *Washington Post,* June 4, 1997.

Wells, H. G. *The Island of Dr. Moreau.* London: Heinemann, 1896.

Westacott, E. *A Century of Vivisection and Antivivisection: A Study of Their Effect upon Science, Medicine and Human Life During the Past Hundred Years.* Ashingdon, Rochford, and Essex, England: The C. W. Daniel Company, 1949.

Wilmut, I., A. E. Schnieke, J. McWhir, et al. "Viable Offspring Derived from Fetal and Adult Mammalian Cells." *Nature* 385 (1997): 810–13.

Wilson, Edward O. *Biophilia: The Human Bond with Other Species.* Cambridge and London: Harvard University Press, 1984.

Wolfle, Thomas. "How Different Species Affect the Relationship." In *The Human/Research Animal Relationship,* edited by Lee Krulisch. Greenbelt, Md.: Scientists Center for Animal Welfare, 1996.

Zawada, W. M., J. B. Cibelli, P. K. Choi, et al. "Somatic Cell Cloned

References

Transgenic Bovine Neurons for Transplantation in Parkinsonian Rats." *Nature Medicine*, vol 4, no. 5 (May 1998).

Zinsser, Hans. *Rats, Lice and History*. London: Macmillan, 1985.

Zurlo, J., D. Rudacille, and A. M. Goldberg. *Animals and Alternatives in Testing: History, Science and Ethics*. New York: Mary Ann Liebert, 1994.

RESOURCES

SCIENTIFIC AND RESEARCH ORGANIZATIONS

The following list of scientific and research organizations and Web sites is provided courtesy of the Animal Welfare Information Center, National Agricultural Library, United States Department of Agriculture.

AAALAC

Association for the Assessment and Accreditation of Laboratory Animal Care. Voluntary accrediting body for demonstrating achievement of certain standards for an animal care and use program.
Albert E. New, Executive Director, 11300 Rockville Pike, Suite 1211, Bethesda, MD 20852–3035; (301) 231-5353.

AALAS

American Association for Laboratory Animal Science. A professional association for veterinarians, animal care workers, managers, and manufacturers involved in laboratory animal science. Publisher of *Laboratory Animal Science and Contemporary Topics.*
Michael Sondag, Executive Director, 70 Timber Creek Drive, Suite 5, Cordova, TN 38018; (901) 754-8620.

AAMC

Association of American Medical Colleges. Through its ad hoc Group for Medical Research Funding, publishes recommendations and guidelines on the use of animals in research.
2450 N Street NW, Washington, DC 20037; (202) 828-0455.

ACLAM

American College of Laboratory Animal Medicine. Certifies veterinarians (Diplomates) who achieve certain standards in laboratory animal medicine.
Charles W. McPherson, Secretary-Treasurer, 200 Summerwinds Drive, Cary, NC 27511; (919) 851-3126.

Resources

AMA

American Medical Association. A professional association of physicians. Published a white paper titled "Use of Animals in Biomedical Research."
515 North State Street, Chicago, IL 60610; (312) 464-5000.

APA

American Psychological Association. An association founded to advance the understanding of basic behavioral principles. Publishes a detailed statement on the care and use of animals titled "Guidelines for Ethical Conduct in the Care and Use of Animals."
750 First Street, NE, Washington, DC 20002–4242; (202) 336-6000.

APHIS

Animal and Plant Health Inspection Service. The division of the U.S. Department of Agriculture that administers the federal Animal Welfare Act.
U.S. Department of Agriculture, Animal and Plant Health Inspection Service, REAC, 6505 Belcrest Road, Room 268-FB, Hyattsville, MD 20782; (301) 436-7833.

APS

American Physiological Society. First scientific society to adopt a written statement on the prevention of cruelty to research animals. Distributes "Guiding Principles in the Care and Use of Animals" to members for signing and posting.
9650 Rockville Pike, Bethesda, MD 20814; (301) 530-7164.

ASLAP

American Society of Laboratory Animal Practitioners. An organization of veterinarians engaged or interested in the practice of laboratory animal medicine.
Brad Godwin, Secretary-Treasurer, 6431 Fannin, Room 132, Houston, TX 77030–1501; (713) 792-5127.

AVMA

American Veterinary Medical Association. A professional association of veterinarians. In 1993, the AVMA published recommended standards for euthanasia procedures that are accepted as national guidelines.
1931 Meacham Road, Schaumburg, IL 60173–4360; (708) 925-8070.

AWI

Animal Welfare Institute. A national organization active in laboratory animal welfare issues. Its sister organization, the Society for Animal Protective Legislation, is a major lobbying force. The AWI encourages laypersons to serve on IACUCs and has a number of publications pertinent to laboratory animal welfare.
Mrs. Christine Stevens, P.O. Box 3650, Washington, DC 20007; (202) 337-2332.

AWIC

Animal Welfare Information Center. The information center of the National Agricultural Library, established as a result of the 1985 amendment to the Animal Welfare Act.
Animal Welfare Information Center, National Agricultural Library, Beltsville, MD 20705; (301) 504-6212.

CAAT

Center for Alternatives to Animal Testing. Established in 1981 to encourage and support the development of nonanimal testing methods. The center supports grants, sponsors symposia, and publishes a variety of materials.
Johns Hopkins School of Public Health, 111 Market Place, Suite 840, Baltimore, MD 21202; (410) 223-1692.

CALAS

Canadian Association of Laboratory Animal Science. A professional association for veterinarians and technicians involved with laboratory animal science.
Donald G. McKay, Executive Director, BioScience Animal Services, M524 Biological Sciences Building, The University of Alberta, T6G 2E9 Canada; (403) 432-5193.

CCAC

Canadian Council on Animal Care. The national body that establishes policy on the care and use of laboratory animals in Canada. Has many useful publications.
1000-151 Slater Street, Ottawa, Ontario, K1P 5H3; (613) 238-4031.

FASEB

Federation of American Societies of Experimental Biology. A federation of leading professional associations, including physiologists, pharmacologists, and other major disciplines involved with animal experiments.
9650 Rockville Pike, Bethesda, MD 20814; (301) 530-7075.

FBR

Foundation for Biomedical Research. A nonprofit educational organization established to inform the American public about the proper and necessary role of animal models in biomedical research and testing.
818 Connecticut Avenue, NW, Suite 303, Washington, DC 20006; (202) 457-0654.

FDA

U.S. Food and Drug Administration, Office of Animal Care and Use. The federal agency responsible for enforcement of the Good Laboratory Practices regulations.
7500 Standish Place, Room 485, Rockville, MD 20855; (301) 295-8798.

IASP

International Association for the Study of Pain. Publishes the journal *Pain* and has developed ethical standards for investigators of experimental pain in animals.
909 NE 43rd Street, Suite 306, Seattle, WA 98105-6020; (206) 547-1703.

ICLAS

International Council for Laboratory Animal Science.
Osmo Hanninen, University of Kuopio, SF-70211, Kuopio 10, Finland.

ILAR

Institute of Laboratory Animal Resources. The part of the National Academy of Sciences that is responsible for laboratory animal issues, as well as for preparing part of the Public Health Service Policy, titled "Guide for the Care and Use of Laboratory Animals."
2101 Constitution Avenue, NW, Washington, DC 20418; (202) 334-2590.

NABR

National Association for Biomedical Research. An association of biomedical facilities concerned with legislation on laboratory animal welfare and with presenting information about the benefits to human health resulting from animal experiments.
Frankie Trull, Executive Director, 818 Connecticut Avenue, NW, Suite 303, Washington, DC 20006; (202) 857-0540.

NIH

National Institutes of Health. A federal agency that disburses funds for biomedical research and sets policy on laboratory animal welfare (Public Health Service Policy).
Office of Animal Care and Use, 9000 Rockville Pike, Bethesda, MD 20892; (301) 496-5793.

NSF

National Science Foundation. A federal agency responsible for disbursement of funds in support of nonbiomedical animal research—that is, zoological and wildlife research.
1800 G. Street NW, Washington, DC 20550; (202) 357-9854.

OPRR

Office for Protection from Research Risks, National Institutes of Health. The office that oversees compliance with the Public Health Service Policy on Humane Care and Use of Laboratory Animals.
9000 Rockville Pike, Building 31, Room 4809, Bethesda, MD 20892; (301) 496-7041.

SCAW

Scientists Center for Animal Welfare. A nonprofit educational organization of scientists that upholds justifiable animal research and conducts programs to help ensure compliance with federal policies, introduction of alternatives where feasible, and sensitivity to humane issues among scientists.
Lee Krulisch, Executive Director, Golden Triangle Building One, 7833 Walker Drive, Suite 340, Greenbelt, MD 20770; (301) 345-3500.

USDA

United States Department of Agriculture. The federal agency responsible for enforcement of the federal Animal Welfare Act. See APHIS.

INTERNET RESOURCES FOR ALTERNATIVES

Alternatives Education Database
http://www.AVAR.org

Alternatives Online
http://embryo.ib.amwaw.edu.pl/~dslado/invitro/Online5a.htm

Alternatives Research and Development Foundation
http://www.aavs.org/htm/ardf.html

Alternatives to Animal Testing Bibliography
http://sis.nlm.nih.gov/altanimals.htm

Alternative to Skin Irritation Testing in Animals
http://www.invitroderm.com

Altweb Alternatives to Animal Testing
http://altweb.jhsph.edu

Animal Welfare Information Center, USDA
http://www.nal.usda.gov/awic

Association of Veterinarians for Animal Rights
http://AVAR.org

Cardiovascular Laboratory Interactive Program, Auburn University
http://www.vetmed.auburn.edu/~branch/cvl.html

Database on In Vitro Toxicology Methods
http://www.invittox.com

Digital Frog International
http://www.digitalfrog.com

European Centre for the Validation of Alternative Methods
http://www.ei.jrc.it/ecvam/

European Research Group for Alternatives in Toxicity Testing
http://embryo.ib.amwaw.edu.pl/~dslado/ergatt/ergatt1.htm

Fund for the Replacement of Animals in Medical Experiments
http://www.frame-uk.demon.co.uk

Humane Innovations and Alternatives Journal
http://www.psyeta.org/hia/index.html

Information on Alternatives Database
http://oslovet.veths.ho/databasesintro.html

Institute for In Vitro Sciences
http://www.iivs.org/

Resources

In Vitro International
http://www.invitrointl.com/

Invittox Online
http://www.invittox.com

Netherlands Centre for Alternatives to Animal Use
http://www.pdk.dgk.ruu.nl/nca.dir/

Norwegian Inventory of Audiovisuals
http://oslovet.veths.no/NORINA

Norwegian Reference Center for Laboratory Animal Science and Alternatives
http://www.oslovet.veths.no/

Procter & Gamble Animal Research and Product Safety
http://www.pg.com/animalternatives/

Searching for Alternatives
http://nersp.nerdc.ufl.edu/iacuc/alternatives.htm

Sources of Support for Research on Alternatives to Animal Use in Research and Testing
http://www.uiowa.edu/~vpr/research/animalt.htm

University of California at Davis Center for Alternatives
http://www.vetmed.ucdavis.edu/Animal_Alternatives/main.htm

ANIMAL AND RESEARCH RESOURCES ON THE WEB

American College of Laboratory Animal Medicine
http://www.aclam.org

Americans for Medical Progress
http://www.ampef.org

Animal Behavior and Welfare Sites
http://www.erols.com/mandtj

Animal Experiments
http://www.api4animals.org/ResearchResources.htm

Resources

Animal and Plant Health Inspection Service, Animal Care
http://www.aphis.usda.gov/reac/

Animal Rights Resource Site
http://arrs.envirolink.org

Animal Welfare Institute
http://www.animalwelfare.com

Applied Research Ethics National Association
http://www.aamc.org/research/primr/arena/

Association for the Assessment & Accreditation of Laboratory Animal Care
http://www.aaalac.org/home.htm

Australian and New Zealand Council for the Care of Animals in Research and Testing
http://www.adelaide.edu/au/ANZCCART

Brain and Mind Magazine
http://www.epub.org.br/cm/

Canadian Council on Animal Care
http://www.ccac.ca

Center for Veterinary Medicine, FDA
http://www.fda.gov/cvm

Community of Science
http://www.cos.com

CRISP Database, NIH Research
gopher://gopher.nih.gov.70/77/gopherlib/indices/crisp/index

Current Research Information System, USDA
http://cristel.nal.usda.gov:8080/

The Electronic Zoo
http://netvet.wustl.edu/ssi.htm-animalresources
http://netvet.wustl.edu/e-zoo.htm-general

Environmental Enhancement of Caged Rhesus Macaques
http://www.primate.wisc.edu/pin/pef/slide/intro.html

Ethics & Animals
http://funnelweb.utcc.uk.edu/~ilsmith/ethics.html

European Convention for the Protection of Vertebrate Animals Used
for Experimental and Other Purposes
http://www.uku.fi/laitoskset/vkek/Sopimus/convention.html

European Society for Laboratory Animal Veterinarians
http://www.eslav.org/eslav_new.htm

Federation of American Societies for Experimental Biology
http://www.faseb.org

National Center for Research Resources
http://www.ncr.nih.gov/

National Library of Medicine
http://www.nlm.hin.gov/

Office for Protection from Research Risks
http://www.nih.gov:80/grants/oprr/library_animal.htm

Primate Enrichment
http://www.brown.edu/Research/Primate/enrich.html

The Whole Brain Atlas
http://www.med.harvard.edu/AANLIB/home.html

U.S. ANIMAL PROTECTION RESOURCES

American Antivivisection Society (AAVS)
Noble Plaza, 801 Old York Road, Suite 204
Jenkintown, PA 19046–1685
Phone: 215-887-0816
Fax: 215-887-2088
http://www.aavs.org/

American Humane Association (AHA)
63 Inverness Drive East
Englewood, CO 80112
Phone: 303-792-9900
Fax: 800-227-4645
http://www.amerhumane.org

American Society for the Prevention of Cruelty to Animals (ASPCA)
424 East 92nd Street
New York, NY 10128
Phone: 212-876-7700
Fax: 212-348-3031

Animal Protection Institute of America (API)
P.O. Box 22505
Sacramento, CA 95820
Phone: 916-731-5521; 800-348-7387
Fax: 916-731-4467

Animal Welfare Institute (AWI)
P.O. Box 3650, Georgetown Station
Washington, DC 20007
Phone: 202-337-2332
Fax: 202-338-9478

Association for Veterinarians for Animal Rights (AVAR)
P.O. Box 208
Davis, CA 95617–0208
Phone: 916-759-8106
Fax: 916-759-8116
http://www.envirolink.org.arrs/avar/avar_www.htm

Feminists for Animal Rights (FAR)
P.O. Box 16425
Chapel Hill, NC 27516
Phone/Fax: 919-286-7333

Humane Society of the United States (HSUS)
2100 L Street NW
Washington, DC 20037
Phone: 202-452-1100
Fax: 202-778-6132
http://www.hsus.org

Medical Research Modernization Committee (MRMC)
P.O. Box 2751, Grand Central Station
New York, NY 10163–2751
Phone: 212-832-3904

National Antivivisection Society (NAVS)
53 West Jackson Boulevard, Suite 1552
Chicago, IL 60604–3795
Phone: 312-427-6065
Fax: 312-427-6524

New England Antivivisection Society (NEAVS)
333 Washington Street, Suite 850
Boston, MA 02108
Phone: 617-523-6020

People for the Ethical Treatment of Animals (PETA)
501 Front Street
Norfolk, VA 23510
Phone: 757-622-PETA
Fax: 757-622-0457
http://envirolink.org/arrs/peta/

Psychologists for the Ethical Treatment of Animals (PSYETA)
P.O. Box 1297
Washington Grove, MD 20880–1297
Phone: 301-963-4751
Fax: 301-963-4751
http://www.psyeta.org

ACKNOWLEDGMENTS

THIS ACCOUNT of the history of the conflict between scientists and antivivisectionists is based on the work of scholars in the humanities, sciences, and social sciences. My aim was not to duplicate their research but to synthesize and interpret their scholarly investigations for a popular audience. The historical passages of this book are reconstructed from biographies, memoirs, and published works of individuals prominent in science and animal protection. Unless noted otherwise, all public statements made by these figures in the book are direct quotes, and are a matter of historical record. The contemporary sections are based on interviews with key figures, the work of scholars who have studied the development of the modern animal rights movement, and my own research and many discussions with people on both sides of this issue during the years I worked as a research writer at the Johns Hopkins Center for Alternatives to Animal Testing. I owe an enormous debt of gratitude to those who have spent all or part of their careers investigating these subjects and working in this field. Their labors facilitated my own investigations, and I encourage readers interested in this subject to seek out the published works listed in the References and to contact the organizations listed in the Resources for further information.

I would particularly like to thank Drs. Alan Goldberg and Joanne Zurlo of the Johns Hopkins Center for Alternatives to Animal Testing. In hiring me as a science writer in 1992, they led me to this material. Their knowledge, guidance, and support, together with my interactions with the members of the Center for Alternatives to Animal Testing Advisory Board and its many supporters during my six-year tenure there, were invaluable in helping me come to terms with the ethical complexity of this debate—an understanding I hope to share with the readers of this book. The Geraldine Dodge Foundation provided financial support for this project, and I am grateful to the foundation and its trustees and executive director for making it possible for me to set down in a book the results of my own years of study on this subject.

Many people read and commented on this manuscript while it was in process. My fellow students in the graduate program in science writing at Johns Hopkins University during the 1997–1998 academic year—John Merryman, Sarah Clemence, Stephanie Renfrew, and Waverly Harrell—provided a great deal of moral support as I wrote the early chapters of the book, and offered perceptive criticism and unfailing encouragement. I owe a special debt to Professor Barbara J. Culliton, Times-Mirror Visiting Professor of Science Writing at Johns Hopkins University, whose support and guidance were particularly helpful during the early stages of writing. I would also like to thank Professor Judith Grossman and Ann Finkbeiner of the Writing Seminars at Johns Hopkins for their support.

Dr. Mark Matfield of the European Biomedical Research Association provided editorial assistance and fact checking of an early draft of the completed manuscript—his comments were invaluable and helped me write a stronger, more balanced history. Rebecca Freeman offered me friendship, support, and comments and criticism on successive drafts. Her comments on the Holocaust and on the subject of women and animals were particularly helpful. Martin L. Stephens, vice president for research issues at the Humane Society of the United States, and Bernard Unti, a friend and postdoctoral student who is working on his own history of the animal protection movement in the United States, also read and commented on the manuscript. I benefited from their comments and suggestions. My fairy godmothers, Carol Ellin and Florence Isaacs, led me to an agent who thought this was an interesting project, and that agent, Linda Konner, not only sold the book but helped keep me sane during its editing.

Acknowledgments

My editor at Farrar, Straus and Giroux, John Glusman, encouraged me to deliver a manuscript that answered not only his own questions, but also those of readers who are neither scientists nor animal protectionists. I am grateful for his persistence and unwillingness to settle for anything less than the very best work I could produce. To all the above friends and colleagues, my deepest gratitude.

Finally, I would like to thank my mother, Jean, and my children, Amelia, Jake, and Sofia, for their love, understanding, and support. They endured three years of upheaval, uncertainty, and financial sacrifice, and I am blessed (and challenged) every day by their wit, courage, and creativity.

February 2000
Deborah Rudacille

INDEX

abortion, 227
Adams, Carol J., 259–60
Adler, Margot, 264
AIDS, 174–75, 224, 225, 229
Alberts, Bruce, 253
Allen, Tim, 194–95
All God's Mistakes (Bosk), 233–34
American Antivivisection Society, 50, 183
American Museum of Natural History, 129–32, 135, 160, 161, 193
Americans for Medical Progress (AMP), 172–77
Ames, Bruce, 165
Anderson, W. French, 212–13, 230–31
Animal Legal Defense Fund (ALDF), 183
Animal Liberation (Singer), 122, 132, 137
Animal Liberation Front (ALF), 4, 135–36, 144, 146–49, 176, 178, 180, 182, 197, 288–91
Animal Rights Crusade, The (Jasper and Nelkin), 140, 142, 143, 203
Animal Rights Movement in America, The (Finsen and Finsen), 140

Animals, Politics and Morality (Garner), 288
Animals (Scientific Procedures) Act (Britain, 1986), 275, 284–85, 286, 287
Animal Welfare Act (Laboratory Animal Welfare Act) (1966), 128–29, 147, 151, 157, 183, 188, 190, 193, 194, 205, 301–2, 310
Animal Welfare Information Center (AWIC), 157, 194–95
Animal Welfare Institute (AWI), 98, 100–1, 106–9, 125, 126, 127, 128, 129, 184
Annas, George J., 236
Antivivisection and Medical Science in Victorian Society (French), 34, 43–44
Aristotle, 20
Arluke, Arnold, 87–88, 309–10, 312
Aronson, Lester R., 129
attention deficit/hyperactivity disorder (ADHD)/attention deficit disorder (ADD), 243–44, 249–50

Backdoor to Eugenics (Duster), 234
Bad Blood (Jones), 96
Baldwin, Alec, 174
Barnard, Neal, 300